HYBRID SHIP HULLS

HYBRID
SHIP HULLS
Engineering Design
Rationales

VLADIMIR M. SHKOLNIKOV
Principal Investigator
Beltran Inc., New York, USA

AMSTERDAM • BOSTON • HEIDELBERG • LONDON
NEW YORK • OXFORD • PARIS • SAN DIEGO
SAN FRANCISCO • SINGAPORE • SYDNEY • TOKYO

Butterworth-Heinemann is an imprint of Elsevier

ELSEVIER

Butterworth Heinemann is an imprint of Elsevier
The Boulevard, Langford Lane, Kidlington, Oxford, OX5 1GB
525 B Street, Suite 1800, San Diego, CA 92101-4495, USA

First published 2014

British Library Cataloguing in Publication Data
A catalogue record for this book is available from the British Library

Library of Congress Cataloging-in-Publication Data
A catalog record for this book is available from the Library of Congress

ISBN: 978-0-128-00861-4

For information on all Butterworth-Heinemann publications
visit our website at **store.elsevier.com**

14   15   16   17   10   9   8   7   6   5   4   3   2   1

Working together
to grow libraries in
developing countries

www.elsevier.com • www.bookaid.org

CONTENTS

PREFACE

The development of novel structural systems is inconceivable without advanced materials aimed at improving service performance for a new demand. As with everything in nature composed of a mixture of materials that work well together, two or more dissimilar material systems may be employed in concert to form a hybrid engineering structure that balances the enhanced service performance of the structure with its feasibility and cost efficiency.

One of the most common structural hybrids being exploited today combines metals with polymer matrix composites. A characteristic example of this is a warship's primarily metal hull with composite structural components. While upgrading functional and operational capabilities, a hybrid hull significantly differs from a conventional monomaterial hull, which is either fully metal or fully composite. The encountered distinctions pertain to the two major factors governing a structure's engineering, material processing and structural behavior. Jointly, these distinctions engender considerable peculiarities of engineering routines that need be ascertained for an effective implementation of the projected hybrid structure.

The present work identifies specific traits intrinsic to hybrid structural systems and outlines design rationales that maximize functional and operational effectiveness of the hybrids. It is primarily dedicated to the advancement of naval structures and is based upon experience acquired over decades of engineering and operation of naval composite and hybrid structures. Aside from a direct military destination, the imparted data might also be useful for a diversity of heavy-duty structural systems throughout non-naval marine, aerospace, automotive, wind-power generation, and other industries striving to maximize service performance, weight efficiency, cost effectiveness, and safety of structural operation.

The book chapters impart diverse aspects of hybrid structure engineering, specifically as follows.

Chapter 1 introduces the grounds for hybrid hull development and peculiarities inherent to hybrid structure engineering. It also provides an overview of the history of composite shipbuilding and summarizes the lessons learned from preceding experience, the rational adaption of which facilitates success in hybrid hull development.

Chapter 2 benchmarks representative examples of warships with hybrid hulls, distinguishing two major categories of composite applications for primarily metal naval vessels: topside structures of surface ships and outboard (light hull) submarine structures.

Chapter 3 enlightens specific facets pertinent to engineering of material-transition structures, hybrid (metal-to-composite) joints in particular, and outlines design rationales for selection of robust and structurally efficient technical solutions.

Chapter 4 presents results of the recent development of an emerging bonded-pinned (Comeld-2) hybrid joining technology that appears superior to the state-of-the-art hybrid joining options for heavy-duty applications available to date. Chapter 4 also delineates economic aspects of hybrid hull concept implementation with utilization of Comeld-2.

Chapter 5 addresses methodological aspects of serviceability evaluation of composite and hybrid structures undergoing long-term changing force-ambient operational exposures.

Chapter 6 identifies and outlines the targets for prospective developments with regard to beneficial utilization of the hybrid structure concept, including those beyond naval shipbuilding.

The appendix of the book presents a few MatLab codes exemplifying evaluation of serviceability of a composite structure undergoing assorted force-temperature exposures.

ACKNOWLEDGMENTS

The author expresses his deep appreciation to Elsevier for its steadfast assistance and guidance during preparation of this manuscript. The author extends his special thanks to Hayley Gray, Senior Acquisitions Editor; Cari Owen, Editorial Project Manager; Hop Wechsler, Permissions Helpdesk Manager; Lisa M. Jones, Project Manager; and Paul Gottehrer, Copy editor.

A prevailing part of the presented data reflects outcomes of the author's investigations primarily dedicated to the advancement of assorted naval platforms utilizing structural composites. These were performed in part in Russia prior the author's move to the U.S. in August 1995 and in part afterward.

The author gratefully acknowledges the sponsorship provided by both the Russian and U.S. governments, as well as the productive cooperation of his colleagues in both countries.

CHAPTER 1

Premises of Hybrid Hull Implementation

1.1 TRENDS IN DEMAND FOR COMPOSITE AND HYBRID STRUCTURES

Development of novel structural systems is inconceivable without advanced materials capable of facilitating service performance related to a new demand. As with everything in nature composed of a mixture of materials that work well together, two or more dissimilar material systems may be employed in concert to form a heterogeneous, hybrid structure that enables a rational balance of enhanced performance with feasibility and cost efficiency of the new structure. One of the most common structural hybrids being exploited combines metals with polymer matrix composites (PMCs).

Structural utilization of PMCs is extensive and rapidly expanding today. This is due to a combination of the structural and physical properties of PMCs that enables substantial advancement of assorted structural systems. A structure's weight reduction allied with the high specific strength of structural PMCs; an opportunity to provide a complex streamlined shape, considerably simplifying employed manufacturing processes; and great corrosion/fouling resistance in a harsh operation environment, allowing for practically effortless maintenance—these and other advantages are driving the exceptional popularity of PMCs for a diversity of structural applications.

Among the major beneficiaries are watercraft, aircraft, and spacecraft; automobiles and other ground vehicles; bridges, causeway floating platforms, and offshore oil/gas rigs; pipelines and pressure tanks; wind turbine blades; and so on. Warships and other naval platforms represent a worthwhile example of the structural hybrids operated on, under, and above the sea surface.

Despite the multiple gains, lack of magnetism was in fact a prime inspiration for the naval application of PMCs, particularly for mine countermeasures vessels (MCMVs). Enhanced stealth performance is another advantage calling for expanded use of PMCs for warships. Not only relatively small and midsize

warships, such as MCMVs and corvettes, which typify full-composite naval vessels, benefit greatly from PMC utilization. Large, primarily metal-hull ships such as destroyers and missile submarines, for which a full-composite hull is impractical, may also be beneficiaries. For instance, a destroyer's superstructure made of a PMC is capable of absorbing electromagnetic emanations from radar and transforming the signature of the vessel, simultaneously significantly reducing her top weight (Arkhipov et al., 2006; Hackett, 2011; Lackey et al., 2006).

In general, such key advantages as weight saving, augmented deadweight-to-displacement ratio, increased speed and/or cruising range, improved stability, corrosion prevention, enhanced propulsion characteristics, and improved signature control may all ensue from implementation of a hybrid hull combatant ship.

> Weight saving, augmented deadweight-to-displacement ratio, increased speed and/or cruising range, improved stability, corrosion prevention, enhanced propulsion characteristics, and improved signature control; all could be facilitated by implementation of the hybrid hull concept for a combatant ship.

Essentially, any structural component of a hybrid hull might be made of structural PMCs, including but not limited to hull shell panels, bulkheads, platforms, the deckhouse, the superstructure, and foundations for machinery and equipment, as well as other heavily loaded ship structures, including rudders and structural components of water jet propulsion systems, such as the outlet, pump housing, housing inlet, and inlet tunnel.

It should be noted that along with the primary structural material, metals and PMCs, an assortment of ancillary materials may be used within a hybrid structure. These include a variety of light-weight core materials pertinent to sandwich panels, rubbers (for acoustical enhancement of structural panels), and ceramics (useful for enhancement of ballistic protection of a structure's panel).

A series of recent patents and technical papers enlighten the hybrid hull notion with regard to the major structural components of a primarily metal naval surface vessel—bow, stern, and midship side panels, as well as topside structures. The following represent an array of related recent patents (Aleshin et al., 2011; Barsoum, 2002, 2005; Critchfield et al., 2003; Kacznelson et al., 2009; Maslich et al., 2009; Shkolnikov, 2011, 2013) and technical papers (Barsoum, 2003, 2009; Bulkin et al., 2011; Critchfield et al., 1991; Horsmon, 2001; Kudrin et al., 2011; Mouritz et al., 2001; Potter, 2003; Shkolnikov et al., 2009).

As for rewarding applications for surface vessels, hybrid structures are also favorable for submarines, particularly in terms of their outboard structural components. The benefits pertaining to PMCs' submarine application include increased sonar efficiency, avoidance of intricate demagnetization procedures relevant to complex-shape structures, and simplified trimming and ballasting operations. For these reasons, a sonar dome, ballast cisterns, superstructures, sail (fairing), fins, propulsors, launch tubes, and hatches are all good candidates for replacement of metal with PMC to enable significant enhancement of a sub's structural and combat efficiency.

Figure 1.1 depicts a generalized hybrid hull architecture applicable to both major categories of naval ships, surface vessels and submarines, for which the hybrid hull option might be superior.

The white areas indicate locations of composite structural components potentially beneficial to the service performance of these metal naval vessels.

Besides technical advantages, a PMC application for a primarily metal vessel may facilitate considerable cost savings. Although a hybrid hull construction itself is typically somewhat more expensive than a conventional monotonous metal hull, the ensuing significant weight savings ultimately provides a noticeable reduction of the ship's construction cost. Resistance to both corrosion and fouling in turn dramatically lowers maintenance expenses, greatly contributing to overall ownership cost savings.

1.2 HYBRID HULL PECULIARITIES

Evidently, a hybrid structure comprises merely metal and composite mono-material components along with a distinctive heterogeneous material-transition structure. For some structural units, such as a hull shell, mono-material components represent the prevailing part of the hybrid structure, while the material transition typically embodies just a hybrid (composite-to-metal) joint. For other parts, such as a ballistic-protection panel or a composite pipeline with a metallic load-sharing liner, the material transition essentially represents the entire hybrid structure. For both these major alternatives, the pursued heterogeneity, while capable of upgrading functional and operational performance, considerably affects both manufacturing technology and structural behavior of a hybrid ship hull, requiring a certain revision of conventional engineering routines, including a structural design optimization, structure analysis and strength reconciliation, and material processing.

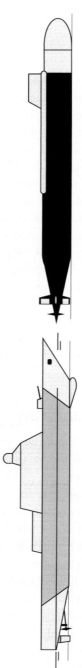

Figure 1.1 Generalized hybrid hull architecture.

First of all, a trade-off study, looking at the feasibility and techno-economic appraisal of the hybrid hull concept implementation, needs be carried out to calculate the scale of the composites that maximizes anticipated technical benefits and cost effectiveness with respect to a particular vessel.

The structural optimization of a hybrid hull is complicated by the multiple design variables inherent to a two-/multi-material structural system. In addition, an extra challenge imposed by the structural heterogeneity is to provide requisite structural integration allied with robust, reliable, and structurally efficient composite-to-metal coupling.

With regard to hybrid structure analysis, distinct properties of utilized dissimilar materials need be taken into account. However, this is not all. The difference in physical properties may initiate derivative interactions between those dissimilar parts inducing additional mechanical stressing and/or other adverse effects. Unequal thermal expansion under altered ambient temperature and galvanic corrosion of the metal part in a seawater environment, attributable to the distinct electrode potential of dissimilar structural components, exemplify that issue.

One more behavioral distinction pertains to a potentially considerable difference in fatigue performance of the dissimilar parts of a material-transition structure. For this reason, a part that has superior load–bearing capability under a short-term loading may manifest inferior performance under long-term operation. This transition can be aggravated by different sensitivities of the dissimilar parts to environmental impacts. Due to these considerations, the weakest link may migrate over the material-transition structure undergoing alternating force-ambient loading exposure during a ship's operation.

One of the principal distinctions that meaningfully affects manufacture of a hybrid structure is simultaneity of composite part processing with processing of the material-transition component. While providing an advantageous opportunity to create complex geometries, practically eliminating multiple assembly and post-processing operations, this calls forth a manufacturing-inclined design—a "design for manufacturability" approach.

The quality control also needs to be upgraded to a broad examination of both the PMC part being formed and the interface thereof with the metal part within the material-transition structure, in lieu of the routine inspection of welding of an ordinary metal hull.

Mostly, said specific traits are interconnected, augmenting the challenge of engineering an effective hybrid structure that properly balances the pros

and cons inherent to a material's service properties and maximizes beneficial operational outcome of the entire hybrid structural system.

1.3 INHERITANCE OF COMPOSITE SHIPBUILDING

Rational adaptation of composite shipbuilding and lessons learned from preceding experience to a large degree promise success in hybrid hull development. With a slight stretch, Noah's ark might be considered composite, as it was built of more than one material, of "gopher wood covered inside and out with pitch." Modern composite shipbuilding, now over a half-century old, implies utilization of PMCs for the principal hull parts of a ship. This is allied with assorted trends in hull structural arrangement, a variety of material compositions and layups, and a range of the material processing techniques. Typically, both hull structure and composite material are formed simultaneously, using the same manufacturing process. Due to this, design for manufacturability is a preferable approach for a composite hull, distinguishing it from the customary design of uniform metal or wood hull structures. While pursued from the commencement of composite shipbuilding, design for manufacturability has grown quite gradually that is mainly due to initial lack of relevant experience. The evolution of the full-composite ship design is well illustrated by the heritage of composite shipbuilding to date.

In fact, the design of midsize glass-fiber-reinforced plastic (GFRP) MCMVs, pioneered by Soviet shipbuilders in the early 1960s, largely replicates conventional metal hull design. In particular, Project 1252—*Izumrud/Zhenya*[1] and Project 1258—*Korund/Yevgenya*, designed and constructed in the Former Soviet Union (FSU) in the 1960s – 1970s, are two MCMV classes of the first generation of full-composite ships ever built.

Hulls of both these MCMV classes are made up of relatively thin solid GFRP skin supported by bidirectional framing that comprises transverse bulkheads, transverse frames, longitudinal stringers, and densely set longitudinal stiffeners. In Figure 1.2 is shown the *Zhenya*'s metal-like composite hull design in transverse section with delineated layout of her midship.

As can be seen, multiple structural members are allotted to support the hull shell and create adequate rigidity and robustness. Similarly to a metal ship hull, spacing of the longitudinal stiffeners typically does not exceed 500 mm. Even the stiffener profile replicates the metal T-beam standard.

---

[1] Hereafter, such notation implies an original Russian vs. NATO classification code.

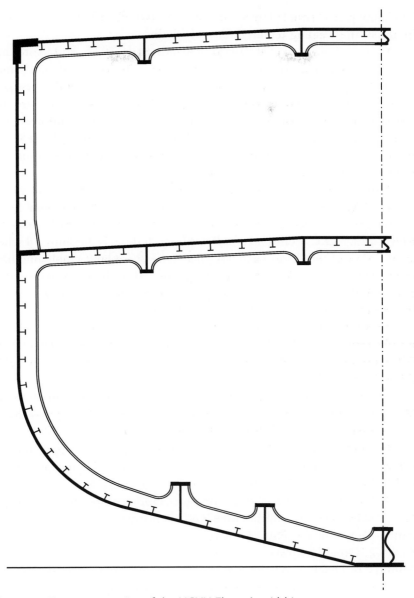

Figure 1.2 Transverse section of the MCMV *Zhenya*'s midship.

The joining of all coupled structural members is fulfilled by bonding patches in lieu of the welding required for metal hulls.

Works of Smirnova et al. (1965), Vaganov et al. (1972), and Skorokhod (2003) represent the primary published sources that describe design principles and impart particular structural details relating to hulls of this first generation of full-composite ships. Utilized material compositions are also imparted. Primarily, this is a polyester resin (PN–609–21M) reinforced with satin fiberglass fabric (T–11–GVS–9). The symmetric and balanced fiber layups $(0°/45°/-45°)$ and $(0°/45°/90°/-45°)$ embody the predominantly utilized laminate schedules.

Both classes of the first-generation composite MCMVs, *Zhenya* and *Yevgenia*, were designed by *Zapadnoe PKB* (currently *Almaz Central Design Bureau*) in St. Petersburg (Leningrad), Russia. *TD "Sredne-Nevsky Sudostroitelny Zavod," LLC*, Pontonny, St. Petersburg, Russia, is the shipyard where all MCMVs of the *Zhenya* and *Yevgenia* classes were built employing hand layup molding of GFRP. Photographs presented in the book authored by Skorokhod (2003) as well as those posted on the Internet—see, e.g., Anon. (n.d.)—illustrate the *Zhenya* and *Yevgenia* MCMVs.

Specifically, the *Zhenya* class was a three-unit series of base 320-ton minesweepers built for the FSU Navy in the late 1960s. These were commissioned in 1966, 1968, and 1969 and served till 1990 (Skorokhod, 2003).

The *Yevgenia* class in turn comprised a 92-unit fleet of minesweepers built for Soviet and foreign navies between 1967 and 1980. They were relatively small 94-ton vessels destined for inshore work.

Each of the *Yevgenia*-class ships served for a long period of operation, typically about 30 years. The latest of these was the RT–71 unit decommissioned in March 2006 (Skorokhod, 2003).

Design and construction of the first-generation composite MCMVs was preceded by extensive collaborative R&D performed by several leading scientific centers of the FSU. Principal was the *Krylov State Research Centre* (*KSRC*). Other key players included the *First Central R&D Institute* of the Ministry of Defense and the *Shipbuilding & Ship Repair Technology Center*, with a great involvement of both design firm and shipyard.

The performed investigations comprised assorted efforts, including devising innovative design solutions suitable for a GFRP ship hull; selection of marine-grade GFRP compositions; working out of material processes relevant to a construction of uniquely large, for that time, structural units within the shipyard environment; and development of structural analysis

and strength reconciliation routines that substantially differed from those used for the conventional metal ship hull.

The analytical studies and design efforts were accompanied by extensive mechanical-environmental testing of material coupons and structure models and prototypes targeted to experimental verification of feasibility, suitability, serviceability, and efficiency of the novel technical solutions as well as relevance and validity of the selected math models. The experimental program included testing to failure of full-scale structural components and hull compartments undergoing static, fatigue, and dynamic loading at varied ambient exposures corresponding to the harsh environment of encountered by naval ships. Ship sea trials concluded the experimental part of the performed R&D program.

Overall, the operational experience of the first generation Soviet MCMVs was quite successful, practically without fault for typically over 25 years of service. Both *Zheya* and *Yevgenia* MCMVs demonstrated sound structural performance and seaworthiness. The principal noteworthy trait of those vessels was the structure's high weight efficiency, allied with the ship hull's low weight-to-displacement ratio. This was 0.281 for the *Zhenya*'s hull and 0.206 for the *Yevgenia*'s hull (Kobylinsky et al., 1997).

Essentially, this weight-related prominence was the outcome of the conventional metal-like hull design embodied by the relatively thin shell supported with densely set framing. The price paid for this advantage was excessive labor related to the fabrication of multiple stiffeners and related vast joining operations that accompanied hull assembly of those vessels. Another negative consequence of copying metal design was manifested by a few instances of local piercing that ensued from occasional rough mooring operations (Yangaev, 2008).

The weight efficiency of the GRFP hulls of following generations of the full-composite MCMVs was typically somewhat lower. The direct follower of the first generation was the *Wilton*, a 450-ton minehunter that plied the seas beginning in 1972. The *Wilton*'s builder was *Vosper Thorneycroft*, Woolston, United Kingdom. The ship was laid down in August 7, 1970; launched on January 18, 1972; commissioned on July 14, 1973; and decommissioned in 1994 (Anon., 2013a).

The *Wilton*'s design was based upon the existing *Ton* class minesweepers built of wood and other non-ferromagnetic materials in the 1950s. The primary material for the *Wilton*'s hull was a composition of glass woven roving with a polyester resin that was to give the vessel a low magnetic signature against the threat of magnetic mines.

The *Wilton*'s hull design, distinct from that of the preceding Soviet composite MCMVs, was notably more composite-inclined. The difference was a relatively thick single solid skin stiffened by scarcely set deep hat-shaped frames. Figure 1.3 delineates a characteristic midship with such a structural arrangement.

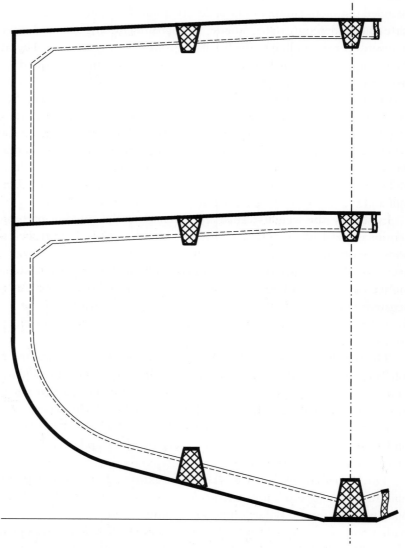

Figure 1.3 Cross-section of *Wilton's* midship: single skin stiffened with hat-shaped frames.

Chalmers et al. (1984) and Smith (1990), along with other experts' publications, consistently illustrate the principles of a composite hull design comprised of hat-stiffened panels. The advantage of these over T-shaped stiffeners lies mainly in the simplicity of the molding operations, with utilization of light-foam core inserts to form a stiffener's profile. Another advantage is the higher structural effectiveness of the hat-shaped frames pertaining to elastic stability, particularly meaningful for in-plane stability of the stiffeners.

The adverse side of these positive features is reduced structural performance of stiffener-to-skin joining by application of secondary-bonded patches especially sensitive to underwater shock exposures. Primarily, diminishing strength is due to the stress concentration at the inner corner of the patch, at the foam-to-shell attachment. To overcome this weakness of mere adhesive bonding it is reinforced with through-skin bolts which ensure the joints' robustness when subjected to underwater shock. The photograph of the *Wilton*'s side view posted by Slemmings (2006) clearly reveals the presence of the bolts reinforcing the adhesive bonding.

It is appropriate to note that although enhancement of the joint resistance to short-term static and dynamic loading was attained, the long-term performance of combined bonded-fastened joints remained practically unchanged, being relatively comparable to the structural performance of plain bonds with the same joint's materials and geometry. A primary reason for such inconsistence was conceivably the adverse influence of the stress concentration related to fastener insertion, which notably affected joint fatigue performance.

> Although resistance to short-term static and dynamic loading increases, the long-term performance of combined bonded-fastened joints stays relatively commensurate with that inherent to plain bonds utilizing the same materials and geometry. A primary envisioned reason for this tendency is an adverse influence of the stress concentration accompanying fastener insertion.

It should be also noted that selection of T-shape framing for the first generation of Soviet MCMVs was partly due to the insufficiency of mere adhesive joining of hat-shaped stiffeners to the hull shell, with regard to ability to withstand dynamic loading exposures.

The progress in closed-mold vacuum-assisted infusion material processing (VIP) attained afterward substantiated significant improvement of the adhesive bonding of the shell and stiffeners (Osborne, 2014). Being applied

simultaneously to the molding of the hull shell and to its supporting hat-shaped stiffeners, VIP enabled uncompromised coupling (Bonanni et al., 2004). This made mechanical fastening of adhesive bonding, such as that employed for the *Wilton*'s hull, unnecessary, which, favorably for the application of hat-shaped stiffeners comparatively to that of T-shaped ones.

The *Wilton* was a coastal MCMV for the Royal Navy, serving as a prototype for the *Hunt* class MCMV with 685-ton displacement, derived from the *Wilton*'s design, construction, and operation experience. The *Hunt* class represented a series of 13 vessels. The first unit commissioned, the *Brecon*, was launched in 1978, and the series was completed by *Vosper Thorneycroft* in 1980 (Anon., 2013b).

Similar to the *Wilton*, the *Hunt* featured the composite-inclined design. The hull had a relatively thick solid laminate skin embodied by glass woven roving and polyester resin. The hull shell was supported with sparsely placed foam-filled hat stiffeners that, along with relatively low-strength, inexpensive fibrous material, caused notable growth of the hull's weight, with a resulting increase in the hull weight-to-displacement ratio to 0.358, as reported by Kobylinsky et al (1997).

To increase the resistance of the frame-to-skin joints to dynamic loading, analogously to the *Wilton*, these combined adhesive bonding patches with mechanical fastening, specifically with thread-cutting screws.

Overall, excluding the drawback pertaining to the relatively weighty hull, the *Wilton-Hunt* design trend was certainly advantageous with respect to both principal characteristics of a composite ship hull, its structural performance, and its manufacturability. While the employed hull arrangement facilitated its requisite robustness and reliability, thus corresponding to MCMV operations, the hull construction cost stayed within a reasonable range. This was mainly due to both the utilization of relatively thick and inexpensive glass fabric reinforcement and the reduced intensity of assembly and skin-frame coupling operations.

The *Sandown* class 484-ton minehunters commissioned in 1989–2001 continued the British *Wilton–Hunt* design tradition with regard to hull structural arrangement. Twelve ships were built for the Royal Navy, and another three were exported to Saudi Arabia. The *Sandown* MCMVs were also built by *Vosper Thorneycroft*, promising the sound structural performance and operational experience manifested by those vessels.

One notable distinction from the *Wilton-Hunt* customary design was the longitudinal orientation of the bottom and main deck structural support that was averred to be more efficient than transverse stiffening in terms of both

weight and construction cost (Shenoi and Wellicome, 2008). It should be noted that mixed stiffener orientation is typically associated with the extra challenge of ensuring the requisite robustness of the bidirectional framing at the ends of intercostal transverse frames.

> Orientation of mixed stiffeners creates an extra challenge to ensuring robust coupling of longitudinal structural members with intercostal transverse frames.

Another major alteration was made with the use of SCRIMP (Seemann composites resin infusion molding process)—a version of conventional VIP (Osborne, 2014), which involved the resin being drawn into a sealed mold under vacuum. In particular, *Vosper Thornycroft*'s SCRIMP-based manufacturing technology was used to construct the entire superstructure, along with some internal structures of the minehunters (Anon., 2012a).

It should be emphasized that closed-mold VIP became imperative for the manufacturing of large FRP structures, such as ship hull components, with long production runs. The primary reasons for this transition were the improved quality of a formed laminate in terms of minimization of the void content; controllability of the fiber volume fraction; superior predictability of weight and mechanical properties of formed PMC; and alleviation of health and safety concerns, as contact with liquid resin was minimized and volatile components of the used resin did not become airborne, being confined by the vacuum bag.

The *Tripartite* class of 605-ton MCMVs, built in France in 1980s, signified one more prominent representative of the British design trend. The design used was a cooperative effort of three nations, the Netherlands, Belgium, and France. The navies of each country operated several ships of the class.

The Italian *Lerici* class of MCMVs represented a further progression of the composite-inclined design toward maximization of hull manufacturability. The first ship of the class was built in 1982 by *Intermarine SpA* for the Marina Militare—the Italian Navy. As stated by the *Intermarine—Rodriquez Cantieri Navali* shipyard (Anon., 2014a), a significant effort was made to develop an innovative structural design, capable of achieving maximum benefit from the intrinsic properties of the GFRP composite, mainly elasticity and flexibility, instead of reproducing a copy of traditional steel or wood ship structures.

The main hull girder in *Intermarine*'s design became fundamentally a heavy single monocoque skin that varied from 25 to 230 mm (Greene,

1999) without any longitudinal or transverse reinforcement other than the main decks and bulkheads, whose strength and stiffness had to be achieved through a significant increase in skin thickness. To achieve this goal, *Intermarine* focused its attention on the dynamic analysis of underwater explosion phenomena of non-contact mines, rejecting various traditional structural solutions and developing a new concept of hull construction.

Intermarine implemented its concept by building a prototype hull compartment, fully representative of the new minehunter's design. The Italian Navy exposed this prototype to severe repetitive underwater explosions with excellent results. A few years later, in 1985, the Italian Navy commissioned four *Lerici* class minehunters.

From then on the design and construction of GFRP minehunters has been *Intermarine*'s core business, and materials and technologies have been continually refined and improved, keeping *Intermarine* on the leading edge of the mine countermeasures market (Anon., 2013c, 2014a).

The *Lerici* class incorporates two subclasses: the first four ships are referred to specifically as the first series of the *Lerici* class, while eight subsequent ships, produced to a slightly modified design, are known as the second series *Lerici* or as the *Gaeta* class (Anon., 2013c).

As of today, the seven navies of Australia, Finland, Italy, Malaysia, Nigeria, Thailand, and the USA have in service MCMVs designed and built by *Intermarine* (Anon., 2013c). The main reason for this is the hulls exceptional robustness is quite suitable for vessels used for mine warefare. Not less important, although all *Intermarine* MCMVs have the same concept of hull construction, their configurations (in terms of mission and propulsion systems) are substantially different: the number of variants implemented for so many different navies is proof of capability of the tailoring the *Lerici* basic design to meet specific operational, logistic, and technical requirements (Anon., 2014a).

In particular, the USS *Osprey* class—the US Navy coastal mine-hunting ships, replicating the *Lerici* design—also has a monocoque shell, with its single skin ranging from 76 mm thick on the topsides to 200 mm thick at the keel (Anon., 1993; Marsh, 2004).

Twelve minehunter ships were built for the US Navy by *Northrop Grumman Ship Systems* (currently *Huntington Ingalls Industries—HII*) in Gulfport, Mississippi, and by *Intermarine* of Savannah, Georgia. These were commissioned in 1993–1999 and decommissioned in 2006–2007.

On the whole, the *Lerici* design trend, atypical for relatively large ships, along with the exceptional hull robustness and damage resistance, brings

substantial advantages to hull manufacturing. This is due to avoidance of extensive framing fabrication and significant simplification of the hull assembly and joining procedures that normally accompany composite ship construction. Together, these factors dramatically reduce the intensity of labor operations and enable a great deal of automation of the hull's construction.

The obvious counterbalancing effect of those advantages is the hull's uniquely high weight, which substantially lowers the vessel's payload capacity, reduces the amount of ammunition carried, increases fuel consumption, and shortens cruising range, significantly damping down the ship's overall performance.

The *Lerici class*'s monocoque design and exceptional hull robustness and damage resistance bring substantial advantages to hull manufacturing due to avoidance of extensive framing fabrication, hull assembly, and joining operations, enabling a great deal of automation of the hull's material layup and processing. The obvious counterbalancing effect of those advantages is the hull's uniquely high weight, which substantially lowers the vessel's payload capacity, reduces the amount of ammunition carried, increases the fuel consumption, and shortens the cruising range, significantly damping down the ship's overall performance.

In addition, the excessive skin thickness compromises the hull's material processing due to encountered overheating inherent to the resin curing, allied with considerable technological stresses and hence a risk of premature delamination of PMC hull structures.

One more major trend in full-composite ship design is a hull comprised of sandwiched panels of relatively thin FRP skins enclosing a light-weight core, typically made of either polymer foam or balsa wood. The sandwiched double skin hull shell is supported by bulkheads, platforms/decks, and sparsely set frames. Due to the sandwiched structure, it also provides heat insulation and absorbs noise.

Although not new for boat-/shipbuilding, this design option was held back from naval applications for a while primarily due to relatively unreliable skin-to-core connection that tends to debonding, especially under shock and/or impact loading routinely intrinsic to warship operation. With progress in the development of polymer adhesives, foams, and VIP techniques this hurdle diminished and use of the sandwich hull design became customary for major naval ship applications (Anon., 2008, 2012b).

The sandwiched double skin of the hull shell, supported by bulkheads, platforms/decks, and sparsely set frames, represents one more design option that is acceptable for naval application, thanks to the development of polymer adhesives, foams, and VIP techniques that ensure the requisite reliability of skin-to-core connection. Due to the sandwiched structure, it also provides heat insulation and absorbs noise.

In particular, the sandwich hull design pertains to several notorious classes of Scandinavian warships, including Sweden's *Visby* class corvette and *Landsort* class MCMV (afterward upgraded and reclassified as the *Koster* class); Denmark's *Flyvefisken* class patrol vessels; and Norway's *Oksoy* and *Alta* classes catamaran MCMVs and *Skjold* class patrol boats, among others (Anon., 2008, 2012c, 2012d, 2014c).

The *Landsort*, the first of the class of seven 360-ton minehunters, was constructed by the Swedish company *Kockums* (formerly *Karlskronavarvet*) for the Swedish Navy (Anon., 2014b). The *Landsort* was commissioned in 1984, followed by construction of the six other minehunters of the class, which were commissioned between 1984 and 1992. The *Landsort*'s hull was made of GRP developed by the Swedish Navy and *Karlskronavarvet*, which is highly durable, easy to repair, and fire- and shock-resistant. The *Landsort* proved to be a robust reliable vessel during her twenty-five-year service in the Swedish Navy prior to upgrade and reclassification to the *Koster* class.

The *Visby*, the first of the *Visby* class of corvettes, one of the most illustrious composite vessels of the present time, was designed by the Swedish Defence Materiel Administration, FMV, and built by *Kockums AB*. Five ships of this class have been completed to date, and two of those were commissioned in December 2009 (Anon., 2013d).

The *Visby*'s hull is designed on stealth principles, with large, flat angled surfaces. Features that are external on conventional ships are concealed within the hull or under specially designed hatches. With a 640-ton displacement, *Visby* is not only one of the largest composite vessels so far built; it is also notable for having a carbon-fiber-reinforced plastic (CFRP) sandwich primary structure—the carbon fiber/vinyl ester skins enclosing a DIAB Divinycell® polyvinyl chloride (PVC) structural foam core (Anon., 2012c).

The *Visby*'s builder, *Kockums AB*, uses stitched nonwoven multiaxials for the carbon skins. Both the fiber architecture and number of plies are varied to correspond to the strength requirements of particular sections of the hull. Panels of up to 60 m^2 are infused using the proprietary *Kockums* vacuum-assisted sandwich infusion (KVASI) system. Each ship is built in three major

sections. After the sections are joined, the complete structure is subjected to a 60 °C post cure (Black, 2003; Marsh, 2004).

Each *Visby* class vessel is propelled by four gas turbines and two diesel engines driving water jets and can exceed 35 knots. The ships can combat mines, submarines, and surface vessels. While conceding that the *Visby* cost one-and-a-half times more to build than a conventional steel corvette, *Kockums* reasons that it has only half the weight and the construction cost difference is recovered in lower fuel costs. Also, much less maintenance is needed over the ship's life, facilitating lowered ship ownership cost over the course of 30 years.

Norway's *Oksoy* and *Alta* are two almost identical classes of 375-ton warfare vessels. The differences are manifested by equipment on the aft decks, the number of sonars, and the length of the superstructure. Corresponding to these distinctions, the *Oksoy* class is used for minehunters, whereas the *Alta* is used for minesweepers. Both *Oksoy* and *Alta* vessels are built by *Kvaerner Mandal* (later *Umoe Mandal*). The vessels are of a 55.2-m catamaran with an air cushion created between the twin hulls as the catamaran moves through the water. The first four ships of the *Oksoy* class were commissioned into the Royal Norwegian Navy in 1994 and 1995. Five *Alta* class ships were commissioned in 1996 and 1997 (Anon., 2012d).

The *Skjold* class patrol boats are conceptually similar to the *Oksoy*/*Alta* class minehunters, but are smaller, having 273-ton displacements. All these air-cushioned catamaran hulls are constructed by *Umoe Mandal* using glass-polyester and glass-vinyl ester laminates over structural polymethacylimide (PMI) foam core. CFRP is also used for high-strength items such as beams, the mast, and support structures. External surfaces, extensively flat and faceted to minimize reflections, are clad in radar absorbent materials. Doors and hatches are flush with the surfaces and windows are flush fitted, without visible coamings. Composites, while simplifying improvement of stealth features, also provide the strength and resilience to withstand wave-induced shock loads, when the combined gas turbine/diesel water jet propulsion system is propelling the craft at speeds of up to 55 knots or more. With her low weight and twin lift fans for surface effect operation, *Skjold* can lift some 70% of her weight out of the water and operate in shallow water (Marsh, 2004).

In the conclusion of this brief survey of the existing trends in composite hull design it should be emphasized that, no matter the particular hull structural arrangement, any implemented design option must provide requisite structural robustness and integrity of the hull for all loading exposures and harsh environmental impacts to facilitate proper warship operation.

Evidently, all the existing hull design configurations delineated above are sufficient for required serviceability pertaining to the whole diversity of operational and specific military loads, which typically comprise seaway normal operational, special operational, and unique military loadings. Each design option features utilization of peculiar material compositions and layups, material processes, and hull assembly techniques, which together to a great degree affect the hull manufacturability, intensity of labor operations, and ultimately the hull construction cost.

The attained hull integrity is largely dependent upon the serviceability of joints between hull sections and other structural components, a critical attribute of a composite hull structure. Joint criticality is heightened by the necessity of compensating for discontinued fiber reinforcement of laminated structural components at the seams with less durable adhesive bonding. This presumes that either mere adhesive bonding or adhesive bonding combined with mechanical fastening or other extra measures would lower the weight efficiency intrinsic to the base hull structures being joined.

In addition, a few more factors downgrade structural performance of joints. The most notorious adverse factors pertinent to joints within large composite structures include:

- Relatively low out-of-plane mechanical properties of laminate PMCs
- Diminished strength and reliability of secondary (post-cured) bonding
- Stress concentration attributable to shape alteration at a joint's region.

In concert, the encountered peculiarities notably complicate provision of robust reliable joining of hull sections and structural components, making this one of the most challenging facets of composite shipbuilding, which greatly affects manufacturability, labor intensity, and the construction cost of a composite hull. For this reason, one of the principal design goals is to reduce the unfavorable influence of joints as much is reasonably possible.

A number of technical measures are routinely employed to meet this demand. One straightforward way is minimization of the presence of joints. The *Lerici* design perfectly exemplifies this notion when contrasted to the first generation of composite MCMVs, which replicated the metal-like conventional design associated with massive joining operations.

Another relevant design rationale is to position seams between hull sections in relatively low-loaded areas of the hull or where some extra material plies are required anyway to locally augment a structure's stiffness. The structural irregularities, such as hull-to-deck connection at the main deck stringer with the hull's sheer strake, the semi-hull seam along the keel line,

or a seam along the hull's bilge, well typify favorable regions for placement of between-sections seams.

Another major trait relates to weight efficiency of the structure, which typically runs counter to hull manufacturability and construction cost. As accentuated above, design configuration significantly affects the hull weight-to-displacement ratio, which in turn alters several key performance parameters of a warship, including her payload and capacity for accommodation of armament and other equipment, as well as speed and cruising range.

Table 1.1 sums up dimensional and displacement characteristics pertaining to MCMV hulls of all four existing major design trends along with relevant hull weight-to-displacement ratios that illustrate this relationship well.

Certainly, the two major performance characteristics—weight efficiency of the hull structure and its manufacturability—should be properly balanced with the imposed design, operational, manufacturing, and cost constraints. For this reason, a trade-off study is exercised at the early design stage in order to select the design-material-technology option that is the best match for manufacturability requirements for hull construction at the assigned shipbuilding facility, construction cost, and serviceability of the particular projected vessel.

1.4 ADVANCED DESIGN-TECHNOLOGY CONCEPTS

Along with the given assortment of fulfilled hull designs, at least one more distinct material-design-technology concept deserves introduction. This concept features structural implication of syntactic foam into PMC layups of major hull components. The main reason for its technical merit is the considerable increase in weight efficiency of such a hull pertaining to relatively large composite vessels.

The concept had been explored under a FSU target R&D program, performed in the 1980s, being specifically destined for full-composite combat ships with displacement ranging from 600 to 1500 tons. The principal goal of the endeavor was to attain the level of weight efficiency of a composite hull close to that of the first-generation composite MCMVs, while enabling substantial reduction of labor intensity by truncating the joining operations.

All significant aspects of composite ship structure engineering were addressed by exercising a systematic approach with regard to the inseparable triad comprising hull structural arrangement, material composition, and

Table 1.1 Principal hull characteristics of full-composite MCMVs

Class	Zhenya[a]	Yevgenia[b]	Wilton[c]	Hunt[d]	Lerici I/II[e]	Osprey[f]	Landsort[g]	Oksoy/Alta[h]
Country	FSU	FSU	UK	UK	Italy	USA	Sweden	Norway
Commission	1966–1969	1967–1985	1973	1980–1989	1985	1993–1999	1980s	1994–1997
Full displacement, ton	320	91.3/96.7[i]	450	750	620/697	804	360	375
Dimensions, m								
Length	42.9	26.1	47	60	50/52.5	57.3	47.5	55.2
Beam	8.25	5.4	8.9	9.8	9.9	11	9.6	13.6
Draft	2.14	1.38	2.6	2.2	2.6	3.7	2.3	2.5
Primary hull configuration	Solid skin supported by bidirectional densely set T-shaped stiffeners	Solid skin supported by sparsely set hat frames and stringers	Solid skin supported by sparsely set hat frames and stringers	Unstiffened solid hull shell	Unstiffened solid hull	Sandwiched hull shell supported by sparsely set hat frames and stringers	
Hull weight/displacement[j]	0.28	0.21	0.27	0.36	0.50–0.60	0.50–0.60	0.3	0.30[k]

[a] Anon. (2013e).
[b] Anon. (2013f).
[c] Anon. (2013a).
[d] Anon. (2013b).
[e] Anon. (2014a).
[f] Anon. (2013g).
[g] Anon. (2014b).
[h] Anon. (2012d).
[i] Revised Project 1258E.
[j] Based on assessments reported by Kobylinsky et al. (1997) unless otherwise noted.
[k] Taby et al. (2001).

manufacturing process. Two distinct design-technology options were specifically addressed. One related to quasi-sandwich material layups comprising plies of fiber material alternating with layers of syntactic foam. The other consisted of a double-bottom hull architecture that was to enable substantial enhancement of longitudinal stiffness of the hull and its robustness under shock exposures, both being critical for naval vessel operation.

The quasi-sandwich panels with syntactic foam possessed notably higher structural efficiency, particularly related to impact resistance, than that of the conventional hull panels, with either stiffened thin-solid skin or a lightweight-cored sandwich structure. Assorted quasi-sandwich layups were explored, targeting maximization of their structural efficiency while alleviating complexity and intensity inherent to labor operations for construction of the first-generation *Zhenya* and *Yevgenia* classes of MCMVs.

Essentially, the novel (at that time) concept fused the *Lerici* monocoque approach with the sandwiched hull architecture intrinsic to the Scandinavian composite warships, such as the *Landsort*, *Visby*, and *Oksoy/Alta* classes, among others. Being optimized, the fiber-reinforced laminate with incorporated syntactic foam allowed for attaining the weight advantage in line with increased structural robustness of the hull.

The double-bottom architecture, in turn, while structurally advantageous, was somewhat intricate with respect to material processing. This was mainly due to the requirement to execute all molding operations from outside the double-bottom structure, to avoid the molding work inside a closed compartment for health and safety reasons.

To overcome this hindrance, several innovative technical solutions were explored. One of those was an arched double-bottom structure. Figure 1.4 delineates this option in conjunction with the quasi-sandwich material layup that is described here, including all engaged structural components: the hull shell, second bottom, stringers, and massive arch-shaped joining enclosures.

The shown design allows for the desired increase in hull rigidity by means of the hull double-bottoming, along with simplification of the hull assembly. The arches formed of syntactic foam reinforced with plies of a fibrous PMC serve a triple function: to augment the hull stiffness, to reduce the effective span between the stringers (thereby increasing shock resistance of the hull), and to provide proper joining of the double-bottom with the base hull and with the stringers.

An array of innovative technical solutions pertaining to the quasi-sandwich material layup, double-bottomed hull design, and relevant manufacturing procedures suitable for construction of large composite

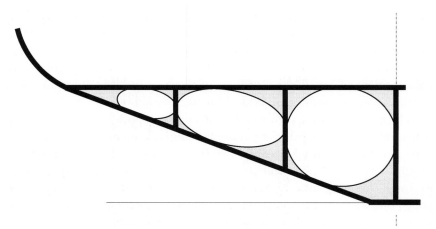

Figure 1.4 Transvers section of arched double-bottom.

vessels was explored. A few nonclassified inventions (Englin et al., 1980; Frolov et al., 1980; Frolov et al., 1995; Shkolnikov et al., 1979; Shkolnikov et al., 1980a, 1980b; Shkolnikov et al., 1982; Shkolnikov et al., 1984; Smirnova et al., 1980) represent the relevant technical solutions for utilizing quasi-sandwich panels and/or double-bottomed hull structures that feature the suggested design options.

Structural superiority of the imparted technical concepts over the existing standards was experimentally verified by implementing an extensive test program comprised of dynamic, static short-term, and fatigue long-term testing to failure of full-scale structural components and double-bottom prototype compartments. A set of design guidelines was developed to support anticipated design efforts aimed at the construction of prospective composite naval vessels utilizing the introduced novel concept. Some of those guidelines are in effect now (e.g., see references OST 5.1001–80 and RD 5.1186–90).

Project 12660—*Rubin/Gorya* class MCMV, that is, the 1150-ton, 67.8-m-long seagoing minesweeper constructed of low–alloy low–magnetic steel (Skorokhod, 2003) was used as a prototype of the targeted new generation of large composite vessels.

Due to the known political and economic changes in Russia in the 1980s – 2000s, the Russian naval forces were downsized several times following the collapse of the FSU. As acknowledged by Barabanov (2004), the

development of minesweepers for the Russian Navy practically ceased after 1991. Only a few minesweepers, designed in the 1970s – 1980s and laid down before the Soviet Union collapsed, have been constructed.

Disappointing for developers and operators, a composite version of the *Rubin/Gorya* MCMV class has never been fulfilled. Nevertheless, many advantageous technical solutions accompanying the implemented R&Ds, partially reflected in inventions (Frolov et al., 1980; Frolov et al., 1995; Shkolnikov et al., 1980a, 1980b; Shkolnikov et al., 1982; Shkolnikov et al., 1984; Smirnova et al., 1980), have not been surpassed to date.

It is appropriate to mention that plans to construct a large full-composite MCMV have recently been revived in Russia. In preparation for that, Russian shipbuilders have successfully molded a quite large full-composite hull suitable for mine countermeasures and patrol vessels (Nekhai, 2011). This is a 70-m-long, 8-m-high hull for a ship of a nearly 1000-ton displacement. A combination of woven glass and carbon cloth has been processed employing a computer-controlled VIP. The hull was formed as a monocoque shell to be outfitted with decks and bulkheads assemblies. This effort is being performed by the *Sredne-Nevsky* shipyard, where the new Project 12700, the *Aleksandrit* harbor class 800-ton minesweeper, designed by the *Almaz Central Design Bureau*, is being constructed for the Russian Navy (Anon., 2013h). Reportedly, it will be a state-of-the-art full-composite ship with characteristics that exceed those of the existing ships of the class. The first ship of the *Aleksandrit* series was laid down on September 22, 2011. Her hull was completed on December 29, 2012 (Anon., 2013i) and her launch initially planned for the mid 2013 is currently rescheduled for June 2014 (Karpenko, 2014).

It should be noted that along with utilization of the advantageous VIP technique to mold the *Aleksandrit*'s hull, its structural arrangement essentially replicates that of the first-generation Soviet composite MCMVs built in the 1960s – 1970s despite all the progress made in composite shipbuilding over the last five decades. As can be seen in photographs, available in Anon., 2011, T-shaped transverse frames and other stiffeners are used as the primary reinforcement of the hull shell. Among probable prime reasons for staying with this traditional design is the necessity to utilize the secondary bonding for coupling of the hull shell with the stiffeners. If this is the case, T-shaped stiffeners do represent a preferable choice, regardless of the general advantages of the hat-shaped stiffeners in terms of weight efficiency, lower labor intensity of both molding and hull assembly operations, and reduced overall construction cost, as they are molded simultaneously with the hull shell.

REFERENCES

Aleshin, M.V., Bulkin, V.A., Gavrilov, V.G., Kozlov, C.V., Ryabkin, V.C., Fedonyuk, N.N., 2011. Composite Superstructure of Seagoing Ship. Patent 2318694, Russia (Надстройка из Полимерного Композиционного Материала Морского Судна).

Anon., 1993. USS Osprey (MHC-51) Minehunter. Available from www.militaryfactory. com/ships/detail.asp?ship_id=USS-Osprey-MHC51.

Anon., 2008. Landsort Class Mine Countermeasures Vessel, Sweden, Naval Technology. Available from http://web.archive.org/web/20080405211435/http://www.naval-technology.com/project_printable.asp?ProjectID=1516.

Anon., 2011. Aft View of Alexandrite MCMV Project 12700. FlotProm, September 28 (Вид в корму корабля ПМО "Александрит" проекта 12700). Available from http://flotprom.ru/media/photo/index.php?PAGE_NAME=detail&SECTION_ID=3759&ELEMENT_ID=92349.

Anon., 2012a. Sandown Class, United Kingdom, Naval Technology. Available from http://www.naval-technology.com/projects/sandown/.

Anon., 2012b. Landsort/Koster Class Minehunter, Sweden, Naval Technology. Available from http://www.naval-technology.com/projects/landsort/.

Anon., 2012c. Visby Class, Sweden. Available from http://www.naval-technology.com/projects/visby/.

Anon., 2012d. Oksoy and Alta Class, Norway. Available from www.naval-technology.com/projects/oksoy/.

Anon., 2013c. Mine Countermeasure Vessels, Intermarine SpA. Available from http://www.intermarine.it/en/products/defence/mcmv.

Anon., 2013a. HMS Wilton (M1116), Wikipedia. Available from http://en.wikipedia.org/wiki/Wilton_class_minesweeper/minehunter.

Anon., 2013b. Hunt-Class mine countermeasures vessel, Wikipedia. Available from http://en.wikipedia.org/wiki/Hunt-class_mine_countermeasures_vessel.

Anon., 2013d. Visby-Class Corvette, Wikipedia. Available from http://en.wikipedia.org/wiki/Visby-class_corvette.

Anon., 2013e. Zhenya Class Minesweeper, Wikipedia. Available from http://en.wikipedia.org/wiki/Zhenya_class_minesweeper.

Anon., 2013f. Yevgenya Class Minesweeper, Wikipedia. Available from http://en.wikipedia.org/wiki/Yevgenya_class_minesweeper.

Anon., 2013g. Osprey Class Coastal Minehunter, Wikipedia. Available from http://en.wikipedia.org/wiki/Osprey_class_coastal_minehunter.

Anon., 2013h. Alexandrite-Class Minesweeper, Wikipedia. Available from http://en.wikipedia.org/wiki/Alexandrit-class_minesweeper.

Anon., 2013i. Sredne-Nevsky Shipyard. Available from http://snsz.ru/?p=669&lang=en.

Anon., 2014a. Lerici-Class Minehunter, Wikipedia. Available from http://en.wikipedia.org/wiki/Lerici-class_minehunter.

Anon., 2014b. Landsort Class MCMV, Wikipedia. Available from http://en.wikipedia.org/wiki/Landsort_class_mine_countermeasures_vessel.

Anon., 2014c, Flyvefisken-Class Patrol Vessel. Available from http://en.wikipedia.org/wiki/Flyvefisken-class_patrol_vessel.

Anon., n.d. MSMVs (Минно-Тральные Корабли). Available from http://sh8146.narod.ru/photoalbummtk.html.

Arkhipov, A.V., Bulkin, V.A., Lazarev, A.M., Ogloblin, Y.F., 2006. Application of PMC for naval surface vessels (effectiveness and perspectives). Strength & Design of Surface Ships of PMC, Works of KSRC 27 (311), 6–22. (Применение Полимерных Композиционных Материалов в Надводном Кораблестроении - Эффективность и Перспективы).

Barabanov, M., 2004. A Survey of Russian Naval Forces: The Surface Fleet in Decline. World Affairs Board. Available from www.worldaffairsboard.com/showthread.php?s=6c6baa6b2664c7adbc1340927892a9ab&t=3630.

Barsoum, R.G.S., 2002. Hybrid Ship Hull. Patent 6386131, USA.

Barsoum, R.G.S., 2003. The best of both worlds: hybrid hulls use composite & steel. AMP-TIAC Quarterly 7 (3), 55–61, Available from http://ammtiac.alionscience.com/pdf/AMPQ7_3ART08.pdf.

Barsoum, R.G.S., 2005. Hybrid Ship Hull. Patent 6941888, USA.

Barsoum, R.G.S., 2009. Application of hybrid hull concept to high speed high performance ships & craft. In: Proceedings of High Performance Marine Vehicles Symposium, ASNE, Linthicum, MD, November 9–10.

Black, S., 2003. Fighting ships augment combat readiness with advanced composites. High-Performance Composites 30–33, September.

Bonanni, D.L., Telegadas, H.K., Caiazzo, A.A., 2004. Testing and analysis of hat-stiffened panel details for navy composite applications. In: The 49th International SAMPE Symposium and Exhibition, CD-ROM Edition, Covina, California.

Chalmers, D.W., Osborn, R.J., Bunney, A., 1984. Hull construction of MCMVs in the United Kingdom. In: Proc. of Int. Symp. Mine Warfare Vessels and Systems, London, UK.

Critchfield, M.O., Morgan, S.L., Potter, P.C., 1991. GRP deckhouse development for naval ships. In: Advances in Marine Structures. Elsevier, London, pp. 372–391.

Critchfield, M.O., Kuo, C.T., Nguyen, L.B., 2003. Hybrid Hull Construction for Marine Vessels. Patent 6505571 B1, USA.

Englin, R.K., Shkolnikov, V.M., Frolov, S.E., Smirnova, M.K., 1980. Method of Joining of Hull Structural Members of GFRP Ships. AC#762322, USSR (Способ Соединения Корпусных Элементов Стеклопластиковых Судов).

Frolov, S.E., Shkolnikov, V.M., Englin, R.K., Smirnova, M.K., 1980. GFRP Panel of Ship Hull. AC#778067, USSR (Перекрытие Корпуса Судна из Стеклопластика).

Frolov, S.E., Shkolnikov, V.M., Smirnova, M.K., Englin, R.K., 1984. Method of Construction of GFRP Ship. AC#1064581, USSR (Способ Постройки Корпуса Судна из Стеклопластика).

Frolov, S.E., Shkolnikov, V.M., Timofeev, B.R., Yarcev, B.A., 1995. Method of Manufacturing of Multi-layer Composite Structures. Patent 2050284, Russia (Способ Изготовления Многослойных Композиционных Конструкций).

Greene, E., 1999. Marine Composites, second ed. Eric Greene Associates, Inc., Annapolis, Maryland, 377pp. Available from http://ericgreeneassociates.com/images/MARINE_COMPOSITES.pdf.

Hackett, J.P., 2011. Composites road to the fleet – a collaborative success story. Special Report 306: Naval Engineering in the 21st Century. The Science and Technology Foundation for Future Naval Fleets, Northrop Grumman Shipbuilding – Gulf Coast, 36pp. Available from http://onlinepubs.trb.org/onlinepubs/nec/61810Hackett.pdf.

Horsmon Jr., A.W., 2001. Lightweight composites for heavy-duty solutions. Marine Technology 38 (2), 112–115.

Kacznelson, L.I., Aleshin, M.V., Bulkin, V.A., Zhegina, V.V., Gorev, Y.N., Sobolevsky, A. A., Samarin, V.S., 2009. Radio-transparent Part of Metal Hull. Patent 2371348, Russia (Конструкция Радиопрозрачной Части Металлического Корпуса).

Kobylinsky, A.V., Paliy, O.M., Spiro, V.E., Fedonyuk, N.N., Fisay, G.V., Frolov, S.E., 1997. Experience and perspectives of polymer matrix composites application for naval surface vessels. In: Proc. of the Second International Conference and Exhibition on Marine Intellectual Technologies (MORINTECH-97), 4, Intellectual Technologies in Applied Research, Russia, SPb, pp. A3.17.1–A3.17.10 (Опыт и Перспективы

Использования Полимерных Композиционных Материалов в Надводном Кораблестроении).

Karpenko, A.V., 2014. Base Minesweeper Project 12700 Alexandrite (Базовый Тральщик Проекта 12700 "Александрит"), Weapons of Fatherland, Domestic Weapons and Military Equipment, June 10. Available from http://bastion-opk.ru/12700-alexandrite/.

Kudrin, M.A., Maslich, E.A., Shaposhnikov, V.M., 2011. Assessment of stress–strain state of long composite superstructures. Works of KSRC 1 (58(342)), 55–58, Russia, SPb, KSRC (in Russian).

Lackey, E., Hutchcraft, E., Vaughan, J., Averill, R., 2006. Zapped electromagnetic radiation and polymeric composites. Composites Manufacturing. May, 7pp.

Marsh, G., 2004. Can composites become serious seagoers? Reinforced Plastics October, 20–24.

Maslich, E.A., Appolonov, E.M., Kudrin, M.A., Fedonyuk, N.N., Shaposhnikov, V.M., 2009. Expansion Joint of Ship Deckhouse Parts. Patent 2402453, Russia (Расширительное Соединение Частей Судовой Надстройки).

Mouritz, A.P., Gellert, E., Burchill, P., Challis, K., 2001. Review of advanced composite structures for naval ships and submarines. Composite Structures 53, 21–42.

Nekhai, O., 2011. Innovations in Russian shipbuilding. Moscow Time, August 19. Available from http://english.ruvr.ru/2011/08/19/54895970.html.

Osborne, T., 2014. An introduction to resin infusion. Reinforced Plastics, 25–29, January/February.

OST 5.1001-80, n.d. GFRP Ships, Structural Joints (Суда Пластмассовые, Детали и Узлы Соединений Корпусных Конструкций).

Potter, P., 2003. Surface ships put composites to work. AMPTIAC Quarterly 7 (3), 55–61, Available from http://ammtiac.alionscience.com/pdf/AMPQ7_3ART05.pdf.

RD 5.1186-90, n.d. Composite Hulls and Hull Structures. Design Rules and Methods of Strength Analysis, USSR (Корпуса и Корпусные Конструкции из Стеклопластика. Правила Проектирования и Методические Указания по Расчетам Прочности).

Shenoi, R.A., Wellicome, J.F. (Eds.), 2008. Composite materials in maritime structures. In: Cambridge Ocean Technology Series 5, vol. 2: Practical Considerations, November, 324p.

Shkolnikov, V.M., 2011. Structural Component for Producing Ship Hulls, Ship Hulls Containing the Same, and Method of Manufacturing the Same. Patent 8020504, USA.

Shkolnikov, V.M., 2013. Material-Transition Structural Component for Producing of Hybrid Ship Hulls, Ship Hulls Containing the Same, and Method of Manufacturing the Same. Patent 8430046, USA.

Shkolnikov, V.M., Englin, R.K., Frolov, S.E., Smirnova, M.K., 1979. Method of Manufacturing of Ship's GFRP Double Bottom. AC#728321, USSR (Способ Изготовления Двойного Дна Корпуса Судна из Стеклопластика).

Shkolnikov, V.M., Englin, R.K., Frolov, S.E., Smirnova, M.K., 1980a. GFRP Ship Hull with Double Bottom. AC#747052, USSR (Корпус Судна из Стеклопластика с Двойным Дном).

Shkolnikov, V.M., Frolov, S.E., Smirnova, M.K., Englin, R.K., 1980b. Method of Joining of Hull Structural Members of GFRP Ships. AC#762322, USSR (Способ Соединения Корпусных Элементов Стеклопластиковых Судов).

Shkolnikov, V.M., Englin, R.K., Frolov, S.E., Smirnova, M.K., 1982. Joining of Hull Longitudinal and Transverse Structural Members of GFRP Ship Hull. AC#904260, USSR (Узел Соединения Продольных и Поперечных Связей Корпуса Судна из Стеклопластика).

Shkolnikov, V.M., Englin, R.K., Frolov, S.E., Smirnova, M.K., 1984. Method of Manufacturing of GFRP Ship Double Bottom. AC#1067739, USSR (Способ Изготовления Двойного Дна Корпуса Судна из Стеклопластика).

Shkolnikov, V.M., Dance, B.G.I., Hostetter, G.J., McNamara, D.K., Pickens, J.R., Turcheck Jr., S.P., 2009. Advanced hybrid joining technology-OMAE2009-79769. In: Proceedings of the ASME 28th International Conference on Ocean, Offshore & Arctic Engineering, OMAE2009, Honolulu, Hawaii, May 31–June 5, 8pp.

Skorokhod, Y.V., 2003. Soviet Mine Countermeasures Vessels (1910–1990), Russia, SPb, KSRC, 230pp (Отечественные Противоминные Корабли (1910–1990)). Available from http://war.kruzzz.com/files/Surface_fleet/P_ships_Skor.pdf.

Slemmings, B., 2006. Former HMS Wilton Side Detail, flickr, July 24. Available from http://www.flickr.com/photos/barryslemmings/sets/72157594209983460/detail/.

Smirnova, M.K., Ivanov, A.P., Sidorin, Y.S., Sokolov, B.P., 1965. Strength of Ship Hull of FRP (Прочность Стеклопластика). Shipbuilding, USSR, Leningrad, 332pp.

Smirnova, M.K., Frolov, S.E., Shkolnikov, V.M., Englin, R.K., 1980. Method of Joining of Crossing Structural Members of GFRP Ship Hull. AC#762321, USSR (Способ Соединения Перекрещивающихся Элементов Набора Судового Корпуса из Стеклопластика).

Smith, C.S., 1990. Design of Marine Structures in Composite Materials. Elsevier Applied Science, New York, NY, 389pp.

Taby, J., Høyning, B., Hjelmseth, A., 2001. GRP in Naval Applications, Possibilities and Production Aspects, RTO-MP-069(II) Norway. Available from http://ftp.rta.nato.int/public/PubFullText/RTO/MP/RTO-MP-069-II/MP-069(II)-(SM1)-11.pdf.

Vaganov, A.M., Kalmychkov, A.P., Freed, M.A., 1972. Design of Hull Structures from Glass-fiber Reinforced Polymers, USSR, Leningrad: Shipbuilding, 137pp. (Проектирование Корпусных Конструкций из Стеклопластика).

Yangaev, M.S., 2008. Inshore Minesweeper Project 1258 "Corundum", Information Portal of Veterans of the 47 BK OVR KTOF (Рейдовый Тральщик Проекта 1258 «Корунд»). Available from www.47br-ovra.com/diviziony-tralshchiki/187-divizion/project-1258.

CHAPTER 2

Existing and Prospective Hybrid Hulls

2.1 COMPOSITE SUPERSTRUCTURES OF HYBRID SURFACE VESSELS

While assorted structures, such as bulkheads, deck panels, foundations, and water-jet housings and inlet tunnels, might be beneficially constructed of polymer matrix composites (PMC), superstructures appear to be the most appealing hull component candidate for a conventional metal surface combat vessel for replacement with PMC. Substantial weight savings, allied with the typically minor contribution of a topside structure to the hull's load bearing under global bending, significant reduction of the warship's signature, and lowering of maintenance expenses are the principal factors accounting for the growing interest in a PMC application for warship topside structures.

The weight reduction translates into greater speed and/or range and payload capacity as well as a generally reduced cost of operation due to lessening of both fuel consumption and maintenance expenses. As this pertains to top structures, the weight reduction also contributes to enhancement of the ship's stability and seaworthiness.

With minor alteration of a structural PMC compound, PMC panels are able to promote acoustic and thermal insulation as well as absorption of electromagnetic radar emanations, without adding any notable weight. This enables a decrease in ship emissions and/or reflections that define her signature, increasing the stealth characteristics of the vessel as a whole (Lackey et al., 2006). Overall, for all these reasons, PMC application for large metal warships is becoming routine practice, especially for superstructures.

> PMC application in large primarily metal warships is becoming routine practice nowadays, especially for topside structures of surface vessels and outboard structures of submarines.

Development of stealth technology for shipbuilding began in the 1970s. The *Sea Shadow* (IX-529), an experimental stealth ship built in 1984 by

Lockheed Martin for the US Navy, represents the first prominent result of those initial efforts. As asserted by Chatterton and Paquette (1994), the *Sea Shadow* represents the application of several advanced ship technologies and ship systems available at that time.

Morylyak (2009) emphasizes that all the means of lowering the ship's signature have been utilized. These encompass proper hull shaping facilitated with small water-plane area twin hull (SWATH) architecture; PMC application; and external radar absorption coating. The *Sea Shadow*'s look, uncommon for warships, was specifically selected to show how a low radar profile might be achieved (Nye, 2012).

Parallel R&D efforts aimed at implementation of PMC superstructures with stealth capabilities have been carried out in several developed countries. In the FSU, the initial efforts were dedicated to development of "Krona"—a glass-fiber-reinforced plastic (GFRP)-based structural material with radar-absorption capabilities. A pilot deckhouse made of Krona was installed on a *Yevgenia* class minesweeper and underwent a trial aimed at verification of structural and radar-absorption capabilities in the Baltic Sea in 1979. All imposed requirements were validated.

In France, analogous R&D efforts have resulted in serial construction of *La Fayette* class frigates—light 3000-ton multi-mission vessels built by DCNS. The *La Fayette* is the world's first operational warship designed from the keel up for stealth and survivability. These vessels feature a modular design that can be readily adapted to the specific requirements of each client navy (Le Lan et al., 1992). Their reduced radar cross section (RCS) is achieved by a very clean superstructure compared to conventional designs, angled sides, and radar absorbent material within balsa–cored sandwich panels, made of GFRP based on polyester resin. Both the deckhouse and deck structure of the *La Fayette* were made of GFRP to reduce weight and provide better fire resistance than aluminum. The core selection arose from balsa's good fire performance relative to charring, low smoke, and toxic byproducts, vital requirements in warship design.

Since the *La Fayette*'s introduction, many modern fighting ships have been designed and built around the world following the same principles of stealth. The hulls of these are furnished, as a rule, with at least a composite superstructure, which is becoming a common attribute of major warships nowadays.

Essentially, the whole assortment of hull design configurations pertaining to full-composite ships, discussed in Chapter 1, is applicable for composite sections and other structural components of a hybrid, primarily metal hull.

Nevertheless, the conventional sandwich structure cored with either light-weight foam or balsa wood is typically employed. The reason for this preference relates to an attempt to minimize the number of stiffeners, thereby reducing the cost, while retaining rigidity of the structure for sensor fit requirements. A few prominent examples of composite superstructure implementation that made it to production, as well as some that did not, are briefly described below.

Hackett (2011) details the history of bringing composite materials to US Navy shipbuilding and the fleet made by *Northrop Grumman Shipbuilding-Gulf Coast* (now *HII*)—one of the main contributors to composite shipbuilding for the US Navy. One example is a success story regarding development of the advanced enclosed mast/sensor (AEM/S) system concept and its facilitation for LPD 17 amphibious assault class ships. Another case study is the DDG 51 Flight IIA composite hangar, which, although it did not make it to the fleet, is of some worth in relation to a lesson learned. The composite high-speed vessel demonstrated the use of composites for the forward one-third of her 88-m-long hull. These large composite structure accomplishments made the next step, that of a composite superstructure with embedded antennas and low observability, an achievable goal. The DDG-1000 class with a composite superstructure became the first class of large US Navy ships so outfitted.

Traditional ship stick masts suffer from sensor blockage from the structure of the mast itself, experience sensor maintenance and preservation issues associated with the corrosive atmosphere in which they operate, and have a high RCS due to the large number of components and the multitude of shapes present. A new generation of mast was required to overcome these deficiencies.

As Hackett (2011) asserts, the composite AEM/S system addresses all of the shortcomings of the legacy mast by enclosing the sensors inside the mast structure and having a flat faceted reflective shell to reduce the RCS of the mast. This protects the sensors inside the mast from the harsh marine environment and corrosive gases of the exhaust plume, and as well as providing safer conditions for performing maintenance on the sensors. The make-up of the composite structure that encloses the radar is tuned to the frequency of the radar behind it, which allows only the desired frequency to enter and exit the composite mast shell, reflecting all other frequencies.

The AEM/S system advanced technology demonstration (ATD) mast being constructed was a 26.5-m-high hexagonal structure that measured 10.7 m across, one of the largest ship composite components ever built

for a ship structure. It was constructed in 1996 and installed in May 1997 aboard the USS *Arthur W. Radford* (DD-968), the *Spruance* class destroyer (Hackett, 2011). The ship's overall dimensions were: length, 172 m; beam, 16.8 m; draft, 8.8 m; and full load displacement, 9200 ton (Anon., 2013a).

The 40-ton structure was fabricated in two halves using SCRIMP. Conventional marine composite materials (E-glass, vinyl ester resin and balsa and foam cores) were utilized throughout the structure. Mechanical bolted joints were placed into both the middle and the base of the structure (Greene, 1999; Mouritz et al., 2001).

The ruggedness of the mast was proven on a couple of unplanned occasions. In February 1999, the *Radford* was involved in a collision with the *Saudi Riyadh*, a 29,260-ton, 200-m-long, roll-on/roll-off container ship, during night operations just off the coast of Norfolk, VA. Neither loss of life nor harm to the AEM/S system mast structure ensued, although the ship as a whole was severely damaged. Also, during its time aboard the *Radford*, the mast survived a nor'easter at sea, again with no damage to the mast structure or antennas (Hackett, 2011).

Overall, the project was deemed a complete success, having exceeded all of its goals. The acquired experience provided a solid base for realization of the following *San Antonio* landing platform dock (LPD-17) class program that required lowering the RCS signature of the amphibious ships. Hackett (2011) acknowledges that along with *Northrop Grumman Ships Systems*, the primary group taking part in construction of the AEM/S system—the largest composite material structure ever installed on US Navy ships—also participating in the development, design, and construction of the AEM/S system were representatives of several US Navy institutions, industry, and academia.

The new AEM/S system mast was a large 28.3-m-high octagonal structure, 10.7 m in diameter, constructed of a multilayer, frequency-selective PMC designed to allow passage of a ship's own sensor frequencies with very low loss while reflecting other frequencies.

The mast's shape was designed to provide a smooth silhouette to reduce RCS. The signature and electromagnetic design requirements, being met, were based on criteria associated with sensor and antenna performance, electromagnetic interference, lightning protection electromagnetic shielding, and electrical bonding and grounding (FAS, 2011). The AEM/S system concept totally modified the topside ship appearance and improved war fighting capability through reduced RCS signature and improved sensor performance, and greatly reduced maintenance costs of the mast and antennas.

The US Navy plans to install the masts in each ship of the *San Antonio* (LPD-17) class, overall length, 208.5 m; waterline beam, 29.5 m; draft, 7 m; and full load displacement, 25,000 tons. Photographs posted at Anon. (2013b) illustrate the LPD-17 appearance with both the forward and after masts. Mouritz et al. (2001) demonstrate assembly of the composite AEM/S system mast onboard an LPD-17 at the *Avondale Shipyard*, part of HII.

The hangar case study targeted to installation of a composite hangar structure onboard the DDG 51 Flight IIA class, although not used, deserves to be mentioned as a lesson learned. The relevant R&D program was initiated in 1991 and was focused on an exploration of the idea that a metal skeleton with composite panels attached was an economical way to manufacture integrated shipboard composites such as a hangar module.

The composite panels were standard size and were adhesively attached to a welded steel frame. As Hackett (2011) describes, the expectation was that this might offer weight savings when attempting to build a large integrated composite structure, when compared to other materials and fabrication methods.

To gauge how nonmagnetic composite panels would integrate into a shipyard facility, the panels were shipped to *Ingalls* for storage and handling. A two-deck-high structure, roughly 6.1 m high by 6.1 m long by 3.05 m wide, fashioned after a section of helicopter hangar, was fabricated, and the interior was outfitted with typical ship systems such as pipe and its hangers, light fixtures, electrical panels, etc. The outfitted hangar module was then subjected to blast resistance testing to determine survivability of the construction and outfitting techniques.

Hackett (2011) admits that the expected weight savings did not materialize because in the hangar concept the beams (frames) were sized to carry the entire load, while the composite plate just kept the weather out. Therefore, the frames were much larger and heavier than if they and the plate were a single composite structure. Hackett (2011) suggests that this experience provided a valuable lesson.

Despite the imposed misconception, the preceding engineering experience with topside ship structures as a whole set the stage for the use of composites for the DDG-1000 destroyer upper-section deckhouse (topside structure), the largest composite structure ever built. The deckhouse was about 900 ton and the hangar was about 200 tons (Lundquist, 2012). The composite topside structures were fabricated by *HII* in Gulfport, Mississippi under the supervision of the ship's prime contractor, *General Dynamics*

(*Bath Iron Works*, Bath, Maine) and then were shipped to Maine for assembly aboard the USS *Zumwalt* (DDG 1000).

LeGault (2010) reports that the DDG-1000's seven-level deckhouse was 48.8 m long by 21.3 m wide by 19.8 m high. The first three levels were constructed of steel, while the upper four levels, embodying the topside structure, were made of the balsa-cored carbon fiber and vinyl ester sandwich panels. The composite topside structure, measuring 39.6 m long by 18.3 m wide by 12.2 m high, contained advanced radar systems and a mission planning/control center.

The ship also featured a composite helicopter hangar built of the same materials. Specifically, the deckhouse had a sandwich construction, featuring balsa between skins made from Toray T700 12K FOE carbon fiber and 510A vinyl ester resin. The T700 fiber was woven into three different fabric patterns, a non-crimp $\pm 45°$ stitched material at a weight of 410 g/m^2, a bonded unidirectional material at 680 g/m^2, and a plain-weave $0°/90°$ fabric at 300 g/m^2. The balsa core used for the majority of the topside sandwich structure was selected for the best combination of mechanical properties, fire-containing capability, and cost value.

The balsa was supplied in three different densities: 0.16, 0.24, and 0.32 g/cm^3; but for panels that required unusually high shear strength, a 0.53 g/cm^3 syntactic foam was used as the core. This was MacroCore®— a high-performance, infusion-ready material with high shear strength, fully isotropic. It was a high-heat resistance macrosphere syntactic product for demanding composite core applications, produced by *Engineered Syntactic Systems* (*ESS*), Attleboro, Massachusetts. Over 99 m^3 of MacroCore was used throughout the deckhouse, in all critical joint areas (Anon., 2014f).

PMC laminate compositions cored with syntactic foam provided much more robustness, with far greater impact and shock resistance properties than those of standard light-weight core materials.

LeGault (2010) and Lundquist (2012) emphasize that the deckhouse superstructure was constructed of flat VIP-treated panels as large as 36.6 m long by 18.3 m wide. Tools are typically coated with a tooling-release agent prior to the lay down of the external reinforcing fabrics, balsa core, and interior fabrics. A woven glass cloth peel ply was also usually placed on the tool or external side of the panel. The cloth was peeled off after infusion and cure, providing a clean, bondable surface for secondary bonding and assembly. The skins, which typically comprised several different types of fabric, were 3.2 mm thick, while the balsa core ranged from a 50.8- to 76.2-mm thickness depending on structural and functional requirements.

For example, the non–crimp, $\pm 45°$ stitched fabric was used in combination with the $0°/90°$ plain-weave fabric in a typical wall panel to create a quasi–isotropic effect. The bonded unidirectional fabric was used in areas and structures where higher stiffness was needed in a single direction. In particular, enlarged beams were used to support a large open area in the helicopter hangar, those beams being made with the unidirectional material. The infused and cured panels were assembled by means of laminate step-downs or "scarves" that facilitate vertical-to-vertical and horizontal-to-horizontal bonds, with wedge blocks between vertical and horizontal panels. Structural putty was used to affix the wedge block between the $90°$ angle formed by the panels, and the block was wrapped in carbon fiber fabric and infused with resin (LeGault, 2010).

The photographs shown by Lundquist (2012) illustrate the fabrication of the composite part of the *Zumwalt*'s deckhouse, whereas Levy (2013) shows the same, integrated with the main metal hull of the destroyer.

The principal pursued benefits were weight reduction and the ability to place systems in the structure during manufacturing, such as antennas, which could not be done with steel or aluminum. In fact, a major feature of the DDG-1000's deckhouse was the antennas (or apertures) which were embedded directly in the structure itself.

The deckhouse structure was also covered with radar-absorbing material. Altogether, with the many ways the DDG-1000 design reduced the signature, this 14,000-plus-ton ship had the RCS of a small fishing boat (Lundquist, 2012).

Although VIP-treated sandwich panels were exclusively used for the *Zumwalt*'s composite deckhouse, pultruded panels, which have the potential to be cheaper, had been considered in the construction of the second ship (LeGault, 2010). Apparently, this option might have been acceptable as it did not intensify the between-panel joining operations anticipated with regard to provision of the requisite robustness and structural reliability of the joints.

Speaking in general, a monocoque deckhouse was also a possibility that would eliminate all between-panel joining associated with intense labor operations and substantial extra weight. For manufacture of a moderate-size topside construction, that appeared to be a viable and cost-efficient option. However, for the *Zumwalt*'s enormous deckhouse this notion was not a practical approach, at least because of the envisioned excessive cost for the mold needed for the monocoque construction.

It is noteworthy that the early development of the stealth technology for the Soviet Navy referred to above has been recently resumed with the new

Russian Project 20380/20382, *Steregushchy/Tigr* class 2200/2250-ton full load displacement corvettes, with hulls measuring 104.5 m in length by 11.0-m beam and 3.7-m draft. The *Steregushchy/Tigr* vessels are designed by the *Almaz Central Marine Design Bureau* and being built at two prominent Russian shipyards, St. Petersburg's *Severnaya Verf* and *Amursky Sudostroitelny Zavod* in Komsomolsk-na-Amure, Khabarovsky Kray, Russia. The three first ships of the class were commissioned in November 2007, October 2011, and May 2013 and are currently in service in the Baltic fleet of the Russian Navy (Anon., 2014a).

The composite deckhouse of the *Steregushchy* goes from side shell to side shell. It comprises solid PMC panels reinforced with sparsely set cap-shape frames and a carling. The utilized material is a flame-retardant laminate PMC combining glass and carbon FRPs. The radar absorption capabilities are provided by special additives applied to the polymer resin and coating (DIMMI, 2011). Along with the deckhouse, the ship is furnished with PMC frame structures supporting the ship's main engines (Bulkin et al., 2013).

Similar to the Western experience, the *Steregushchy*'s superstructure provides significant reduction of the ship's radar signature, thanks to the chosen hull architecture and fire-resistant, radar-absorbent GFRP applied in the tophamper design. Utilizing PMC for the superstructure also allows for a substantial reduction of the top-weight, which is beneficial to several key performance parameters of the vessel. Novel technical solutions were widely used during construction of the ships, including 21 patents and 14 new computer programs (Anon., 2012a).

The composite superstructures were manufactured at the *Sredne-Nevsky Shipyard* and shipped to a ship assembly site either by the Neva River in the case of the *Severnaya Verf* (Atalex, 2012), or by sea in the case of the *Steregushchy/Tigr* class corvette built at the *Amursky Sudostroitelny Zavod Shipyard* (Bmpd, 2012).

Methodological support for the engineering of the composite superstructure was provided by *KSRC*. The technical issues addressed are reflected in several published papers (Appolonov et al., 2002, 2006, 2011; Arkhipov et al., 2006; Bulkin et al., 2006; Fedonyuk, 2006; Kudrin et al., 2011). These papers essentially address all major aspects of the innovative structural design and analysis of the composite deckhouse.

In particular, they describe the structural contribution of the superstructure to the hull's resistance to global bending; the structural efficiency of the metal-to-composite joint between the hull and superstructure; and the evaluation of fatigue performance of the metal hull with the composite superstructure.

Technical rationales and novel structural concepts derived from the analytical and experimental investigations were incorporated into the design of the composite superstructure. The principle technical solutions utilized are reflected in a series of related Russian patents, including ##2318694; 2333131; 2371348; 2402453; and 2429155, authored by Appolonov et al. (2006, 2011).

Bulkin et al. (2011) summarize an experience of the *Steregushchy*'s superstructure operation. The lessons learned are reflected in the revised composite superstructure design for the second and following vessels of the class.

The new Project 22350 class stealth frigate, inaugurated by the *Admiral Gorshkov*, was also furnished with a solid PMC deckhouse (Korablev, 2010). The new frigates were designed by the *Severnoye Design Bureau* and constructed at the *Severnaya Verf*, both in St. Petersburg, Russia. The principal dimensions of the new frigate were length 130 m, beam 16 m, draught 4.5 m, and full displacement 4500 ton. The first frigate of the class was launched on October 29, 2010, and underwent sea trials in the Barents Sea in November 2012. The flagship of the class was planned for commission in November 2013, after which she was to join the 14th Anti-Submarine Warfare Brigade of Russia's Northern Fleet (Anon., 2014b; Mikhailov, 2012).

Concluding the overview of prior experience pertaining to the implementation of the hybrid hull concept for naval surface vessels, it should be affirmed once again that composite structures have become a common attribute of the primarily metal hulls of modern warships. Composite superstructures represent the most popular addition to metal hulls due to the combination of physical and structural properties of PMCs that are beneficial to a naval application.

Essentially, two distinct structural configurations are being used for composite superstructures. These are either solid shell structures reinforced with sparsely set hat-shaped frames or sandwich, primarily light-weight-cored-shell sandwich structures. Both these configurations are well known based on preceding design, building, and operation experience relevant to full-composite ships.

Composite superstructures represent the most popular addition to metal hulls due to the beneficial combination of the physical and structural properties of PMCs for naval application. Essentially, just two distinct structural configurations are used for the composite superstructures, which are either solid shell structures, reinforced with sparsely set frames, or sandwich, primarily light-weight-cored shell structures.

The material compositions are widely varied. Typically, there is a thermo-set composite laminate based on either glass or carbon FRP, or a combination of these. The polymer foam, balsa wood, and syntactic foam represent typical core material options for sandwich panels and hat-shaped frames, all being customary constituents of sandwich structures used for full-composite ships. A monocoque shell structure or an assemblage of separate panels represent the currently employed construction options for composite superstructures.

As a matter of fact, there is no definitive design-technology solution for a composite superstructure. Rather, the most suitable and beneficial option should be selected with regard to a given set of operational, functional, and combat requirements as well as the imposed manufacturing and/or budgetary constraints based on specific techno-economic grounds.

2.2 COMPOSITE OUTBOARD SUBMARINE STRUCTURES

Submarine outboard light hull structures—i.e., those that play no role in maintaining atmospheric pressure inside the structure, normally intrinsic to the pressure hull—represent another major category of ship structure for which a PMC application is suitable and rewarding.

> Submarine outboard light hull structures represent another major category of ship structures for which a PMC application is suitable and rewarding.

PMCs have been in use for outer submarine structures since the early 1950s. The precedent was set by a fairwater top structure installed on the USS *Halfbeak* (Anon., 2013c; Greene, 2006). Later, PMCs were extensively used for light hulls of small submarines and deep-submergence vehicles (DSVs), providing significant enhancement in the capabilities of those vehicles. The outer hulls of the Soviet midget sub Project-865 *Piranha/Losos* class and DSV Project-1096 *Poisk-6* vessels exemplify characteristic applications of PMCs for this category of naval hybrid structures.

The *Piranha* class vessels, special mission midget subs, were designed by the submarine design bureau *SPMBM Malachite* and constructed at the *Novo-Admiralty Shipyard* in St. Petersburg, Russia, in the 1980s. *Piranha*'s displacements were just 218 tons (surfaced) and 390 tons (submerged) and her dimensions were $28.2 \times 4.8 \times 5.1$ m for length, beam, and draft, respectively. The operational depth of hold was 240 m and the maximum depth

was 300 m (Anon., 2000). Two *Piranha* class subs were in service in from 1988 to 1997 (Anon., 2012b).

The *Piranha*'s hull was made of a titanium alloy outfitted with GFRP bean-shaped side panels enclosing ballast tanks. Structurally, the GFRP panels replicated a conventional metal-like design with a relatively thin shell stiffened by transverse frames.

Utilization of the nonmagnetic structural materials greatly reduced the signature and effectiveness of rival magnetic-anomaly detectors or magnetic limpet mines for this type of sub. More detail specification on the *Piranha*'s hull design, her structural arrangement, and her appearance can be found in the references (Anon., 2000, 2010).

The deep-sea vehicles of DSV *Poisk-6*, Project 1906, dedicated to searching and exploring operations for the Soviet Navy at an ocean depth of up to 6000 m, were designed jointly by two submarine design bureaus, *LMPB Rubin* and *CPB Volna* (now SPMBM "*Malachite*") and constructed by the *Novo-Admiralty Shipyard* in 1975 to 1979 in St. Petersburg, Russia. In December 1979, the first and only DSV *Poisk-6* vehicle was launched and underwent multiple sea trials on the Black Sea and Pacific Ocean, lasting for years, till 1987 (DIMMI, 2012).

The AGA-7's length, beam, and draft were 29.0 × 6.5 × 8.2 m, respectively. The vessel was equipped with an entire GFRP light hull, which enclosed ballast tanks filled with gasoline to provide the requisite buoyancy. Similarly to the *Piranha* class subs, the AGA-7's light hull embodied a conventional metal-like design with a relatively thin shell, stiffened by frames. The appearance of the surfaced AGA-7 and her hull cut-out are available and may be seen in DIMMI (2012).

An opportunity to employ PMCs for outboard structures of large subs was intensively pursued by both the Western world and the FSU for a long time, starting during the Cold War. However, practical use of composites for this submarine category remains insignificant, in spite of the encouraging results of target R&Ds, the sound design and construction experience attained to date, and the excellent outcome of operational performance.

The scale of possible introduction of a composite structure into a sub's light hull mainly depends on her architecture, which is distinguished by three distinctive configurations of the hull: single, double, and intermediate "one-and-a-half" constructions.

As known, American (as well as most other western) submarines, including the latest active US Navy *Virginia* class nuclear-powered attack sub, primarily utilize single-hull architecture. This implies a presence of just a few

light outer structures, such as bow, stern, sail (fairing), and superstructure (if any), as well as various hydrodynamic control fins and propulsors. The outer single-hull structures typically house main ballast tanks and provide the sub with a streamline shape, while the main cylindrical part of the pressure hull beyond the light hull sections has only a sound-absorbing perforated rubber or anechoic plating layer.

Contrary to Western custom, double-hull architecture is common for Soviet/Russian submarines. This pertains to all four generations of nuclear subs, distinguished by the technology implemented for their main systems: nuclear reactors, machinery, weapons, sonar, and electronic equipment, among others (Anon., 2014c; Dronov, 2002). The newest Project 935/955—*Borei* class (also known as the *Dolgorukiy* class after the name of the lead vessel) that represents the fourth generation of Russian nuclear ballistic missile submarines is included.

As is common when dealing with a complex engineering system, each design option has certain pros and cons. Pertaining to double-hull architecture, rewarding traits relate to:

- An opportunity to mount equipment outside the pressure hull, allowing for lessening of its dimensions and thereby reducing the sub's material consumption as well as her construction and maintenance costs
- The possibility of placing framing of the pressure hull externally, saving space inside the pressure hull and thus enabling its further downsizing and weight saving
- Increased operational safety associated with enhanced damage stability of the sub's pressure hull due to absorption of some impact energy being applied by her light hull, not compromising the sub's integrity
- Improved control of internally induced noise by the light and pressure hull decoupling that allows for enhancement of the stealth performance and/or simplifies internal layout and equipment mounting
- An opportunity to place a sound-absorbing rubber/anechoic layer within the PMC structural shell of the light hull, thus increasing mounting reliability, which can be a troubling issue (Hooper, 2010).

The negative impact of double-hulling typically relates to the sub's enhanced overall dimensions and the associated increases in drag, energy consumption, and signature.

It is noteworthy that Russian designers, despite the positive experience with double-hull operation, have undertaken a step toward partially single-hull architecture, i.e., the one-and-a-half hull design. Specifically, subs of Project 885, involving construction of *Yasen/Graney* (*Severodvinsk*) class nuclear multipurpose attack submarines, which, along with *Borei* class vessels, embody

the newest, fourth generation of Russian subs, have, in addition to the light components normally intrinsic to the single-hull architecture, an extended bow section and outboard structures behind the central compartments, including a block with vertical launch tubes for cruise missiles. See the *Yasen/Graney* sub's cut-out view, illustrating her structural arrangement in Anon. (n.d.).

While the Russians make a move toward the single-hull design, American designers are considering a transition to a double-hulled option for future submarines, to improve their payload capacity, stealth, and range (Anon., 2014c). The described changeovers perfectly illustrate the notion of the great design variability of a complex engineering system such as a submarine, corresponding to given sets of design and operational requirements. Ultimately, selection is driven by viable technical solutions that satisfy the given requirements in the most effective and economical way.

Essentially, any sub's outer structure might be made of PMC regardless of its particular destination and dimensions, excluding the hull's keel, a PMC implication which for large subs is typically impractical and/or unrewarding. Besides nose sonar domes, which are nearly obligatory, indubitable application targets include: top structures, the sail, external ballast tanks, the foundation for sonars, fins, propulsors, and the tail cone, as well as hatches for assorted launch tubes.

> Essentially, any sub's outer structure may be made of PMC, regardless of its particular destination and dimensions, excluding the hull's keel of a large sub, for which PMC application is impractical and/or unrewarding.

As mentioned above, along with benefits similar to those of surface ships, a submarine PMC application enables increased sonar efficiency, avoidance of intricate demagnetization procedures pertinent to noncircular structures such as fins and hatches, and simplification of the trimming and ballasting operations. Together, these traits are capable of providing significant enhancement of a sub's structural, stealth, and overall combat performance.

Certainly, PMC structural components of a sub's light hull must be sufficiently robust to withstand relevant operational exposure for an assigned service life. Like any hull structure, outboard submarine structures undergo assorted operational loading which, along with loads common for both surface ships and submarines, comprise specific submarine-related loading exposures primarily relevant to surfacing operations. These include:

- Blowing of ballast tanks
- Surfacing from under an ice field, primarily challenging for the submarine's sail structure

- Emergency surfacing allied with a cantilever bending of the bow shell being filled with water.

The iconic photo of the *Birmingham* (SSN-695), a *Los Angeles* class sub executing an emergency ascent demonstration during her sea trials (Anon., 1978), perfectly illustrates this load case.

All structural design trends with respect to the above-described surface vessels are applicable to outboard submarine structures as well, excluding those utilizing a relatively soft (polymer foam or balsa wood) core with sandwiched panels and hat-shaped frames.

In particular, an outboard light hull PMC component might consist of either stiffened or unstiffened structures, to which a skin laminate might be added, with distinct material plies to provide multi-functionality of the light hull.

The assigned functions, along with customary structural and surfacing-provision capabilities, typically include sound-absorbing/anechoic and thermal insulation as well as a provision for the extra buoyancy of topside structures. For instance, a PMC laminate being added with an elastomeric/rubber layer could be tailored to provide the requisite structural performance in combination with absorption of noise radiation caused by structural and waterborne vibrations.

A composite bow sonar dome (Anon., 2014d) and a sail cusp (Anon., 2012c), being produced by *Goodrich* in its Engineered Polymer Products (EPPs) facility in Jacksonville, Florida, for the *Virginia* class 115-m-long nuclear fast-attack submarines, exemplifies the solid unstiffened shell design. The dome is a 22,148-kg, 6.4-m-long, 7.9-m-diameter (at the attachment end) hydrodynamically shaped composite structure that houses and protects the sonar transducer sphere as well as composite towed array fairings, the high-frequency sonar chin array, and pylon fairings (Anon., 2014d; Gardiner, 2012). A thick, single-piece rubber boot is bonded to the dome as an acoustic performance enhancement. Minimal sound energy absorption and reflection properties inherent to the rubber material enhance submarine detection capability. The *Goodrich*'s bow sonar dome shown at Anon. (2014d) illustrates this structure ready for shipment and following installation on a *Virginia* class sub.

The *Goodrich* composite sail cusp in turn is a single-piece composite fairing to be attached to the hull and lower leading edge of the submarine sail (the vertical fin on top of the hull). The complex double-curve shape of this lightweight structure allows for smooth laminar flow of water over its surface, thereby improving the hydrodynamic performance of the submarine (Anon., 2012c).

The original cusp was made from numerous steel components with stiff-eners, fitted together and welded, then filled with syntactic foam to inhibit corrosion and finally welded to the sail and hull structures. This method was material- and labor-intensive due to the sail cusp's complex double curva-ture and the number of parts required to fit and attach the cusp to the hull. Because the steel cusp was welded in place, it was not readily removable for maintenance. The composite sail cusp offers a corrosion-resistant, syntactic-filled structure, with inner and outer composite skins bolted to the sail and hull. Reportedly, the cusp is 2268 kg lighter and $150,000 less expensive than the steel version to manufacture and provides an estimated $20,000 in savings per periodic sail maintenance (Gardiner, 2012).

Per Gardiner's (2012) introduction, *General Dynamics Electric Boat* of Groton, Connecticut, is looking at what role composites can play in the US Navy's upcoming replacement of the 14 aging *Ohio*-class ballistic missile submarines with 12 new 170.7-m-long, 13.1-m-diameter vessels, targeting 2019 for lead ship construction (Anon., 2014e). Based on *Virginia*-class suc-cesses with unstiffened solid shell composite structures, these are likely to be important in achieving the recently announced reduced cost goal of $4.9 billion for the *Ohio*-class replacement.

The earlier Soviet experience features the stiffened panel design for a sub's light hull application beyond sonar dome. This is a sort of tubbing panel conventional for the mining industry. A PMC tubbing panel unites the light hull's solid laminate shell with its I-shape framing, allowing for elimination of the labor-intensive T-joining operations usually accompanying the man-ufacture of stiffened hull panels.

In particular, the midget *Piranha* subs and the DSV *Poisk-6* type referred to above were both furnished with light hulls made up of tubbing panels. Also, one of the Soviet Project 627A—*Kit/November* class subs, which represents the first generation of the Soviet nuclear-powered attack submarines built in between 1957 and 1963 and serving from 1958 through 1991 (Anon., 2013d), had been equipped with such PMC tubbing light hull structures.

Two PMC superstructure panels, topping the pressure hull, and eight analogue side panels (four on each side of the sub), forming ballast tanks, were installed, replacing the conventional metal panels. Together, the installed PMC panels constituted the major part of the sub's light hull.

All the PMC panels had been outfitted with metal skirts, using a com-bined bonded-bolted joint, to be connected to the adjacent metal shell employing conventional metal-metal welding. Also, the framing of the PMC panels was coupled with the outer ring frames of the metal pressure

hull via evenly set L-shaped metallic struts bolted to the PMC frames and welded to the metal ones.

Analogously to hulls of the first MCMVs, *Zhenya* and *Yevgenia*, the marine-grade GFRP composition was utilized, comprised of a polyester resin (PN-609-21M) reinforced with fiberglass satin fabric (T-11-GVS-9). Symmetric and balanced fiber layups (0°/45°/−45°) were primarily used.

Overall, this was a successful, nearly flaw-free experience, except for accidental damage of one of the side PMC panels revealed during the hull inspection accompanying the sub's planned docking in 1973. This was roughly a 0.5-m-long rupture of one of the tubbing frames, accompanied by buckling of the two adjacent metal struts supporting the PMC panel. Supposedly the acquired damage was caused by an improper mooring operation. The GFRP shell itself did not suffer any notable damage and did not lose its watertightness, but rather was just slightly deflected inward due to the residual deformation of the buckled metal struts. The damage was promptly fixed and no other harm and/or failure has been reported regarding the service performance of the light hull PMC panels of the sub.

It is fair to assume in this regard that maintenance of the integrity of the light hull shell is due to the relatively low modulus of elasticity of the utilized GFRP. In the same accident, a customary steel shell would probably experience a breach and/or significant plastic deflection, threatening the safety of the sub's operation.

To address the mooring-related collisions, an innovative configuration of an outer composite shell was then conceptualized. Essentially this is a sort of fender device incorporated into the composite side panels of the light hull at the waterline region.

The other PMC applications for submarine outboard structures have also resulted in largely positive outcomes. Mostly, these relate to multiple installations of PMC nose sonar domes and sail structures in several classes of submarines, in both the West and the FSU.

Accidents which occasionally occur with submarines, involving the nose sonar dome in particular, typically have nothing to do with the material used for the wrecked structure. The incident that occurred in 2005 involving the USS *San Francisco,* a *Los Angeles* class sub, which, while traveling in excess of 33 knots struck an underwater mountain 360 miles south of Guam, exemplifies such regrettable occasions (Anon., 2005; Mount, 2005).

With regard to prospective potentially rewarding applications of PMCs for submarine outboard structures beyond sonar domes, at least three targets should be pursued. One is lowering the acoustic signature of double-hulled

subs by introducing vibration–absorbing structural connectors between pressure and light hulls, in addition to the conventional special coating and sound–absorbing layers of the composite shell.

Another is adaptation of PMC laminate compositions combined with syntactic foam, which could be quasi–sandwich (discussed in Section 1.4 with respect to surface vessels); uniform syntactic foam lump wrapped with plies of a laminar PMC forming the structure's skin; or nonuniform fiber-reinforced syntactic foam lump. Such material compositions are especially suitable and beneficial for subs' fins, hatches, sails, and superstructures; the primary advantage of these is the opportunity to combine high structural performance with provision of extra buoyancy of topside structures, favorable for trimming and ballasting operations.

One more emerging category of the use of PMCs for sub light hull components pertains to structural support of nose sonars, which, along with structural robustness, may require acoustic disconnection of the sonar from the hull in order to provide proper sonar effectiveness.

All these prospective targets appear to be achievable by applying structural PMCs, which are proven to be capable of supplying the requisite robustness and serviceability for long–term operation along with added qualities beneficial to combat efficiency. Overall, utilization of PMCs for outboard structures represents a source for significant improvement of the structural, stealth, and combat performance of submarines.

> Utilization of PMCs for outboard structures represents a source of significant improvement in the structural, stealth, and combat performance of submarines.

The long experience of successful operations with PMC light hull structures has convincingly validated this optimistic expectation.

REFERENCES

Anon., 1978. Birmingham (SSN-695), NavSource Online: Submarine Photo Archive. Available from http://www.navsource.org/archives/08/08695.htm.

Anon., 2000. Project 865 Piranya Losos Class. Federation of American Scientists, September 17. Available from http://www.fas.org/man/dod-101/sys/ship/row/rus/865.htm.

Anon., 2005. USS San Francisco Investigation Completed. Story Number: NNS050509-14, America's Navy, May 9. Available from http://www.navy.mil/submit/display.asp?story_id=18257.

Anon., 2010. Project 865 "Piranha" (NATO – "Losos"), Deep Storm (Проект 865 "Пиранья"). Available from http://www.deepstorm.ru/.

Anon., 2012a. Russian Navy to Receive Corvette Boiky by Year End. RusNavy.com, 16 November. Available from http://rusnavy.com/news/navy/index.php?ELEMENT_ID=16470.

Anon., 2012b. Russian Navy, Submarine Piranha (Project 865) (Военно-Морской Флот России, Подводная лодка Пиранья (Проект 865)). Available from www.navy.su/navyfrog/sub/piranya/index.html.

Anon., 2012c. Goodrich Delivers First Composite Sail Cusp for Virginia Class Submarine. Reinforced Plasics.com, News, June 12. Available from http://www.reinforcedplastics.com/view/26255/goodrich-delivers-first-composite-sail-cusp-for-virginia-class-submarine/.

Anon., 2013a. USS Arthur W. Radford (DD-968), October 22. Available from http://en.wikipedia.org/wiki/USS_Arthur_W._Radford_(DD-968).

Anon., 2013b. San Antonio Class – Amphibious Transport Dock. Available from http://www.military-today.com/navy/san_antonio_class.htm.

Anon., 2013c. USS Halfbeak (SS-352), Wikipedia. Available from http://en.wikipedia.org/wiki/USS_Halfbeak_(SS-352).

Anon., 2013d. November Class Submarine, Wikipedia. Available from http://en.wikipedia.org/wiki/November_class_submarine.

Anon., 2014a. Steregushchy-Class Corvette, Wikipedia. Available from http://en.wikipedia.org/wiki/Steregushchy-class_corvette.

Anon., 2014b. Admiral Gorshkov-Class Frigate, Wikipedia. Available from http://en.wikipedia.org/wiki/Admiral_Gorshkov-class_frigate.

Anon., 2014c. Submarine, Wikipedia. Available from http://en.wikipedia.org/wiki/Submarine.

Anon., 2014d. Marine Composite Structures, Products, UTC Aerospace Systems. Available from http://utcaerospacesystems.com/cap/products/Pages/marine-composite-structures.aspx.

Anon., 2014e. Ohio Replacement Submarine. Available from http://en.wikipedia.org/wiki/Ohio_Replacement_Submarine.

Anon., 2014f. USS Zumwalt (DDG 1000) Guided Missile Destroyer Is State-of-the Art, Engineered Syntactic Systems. Available from http://www.esyntactic.com/applications/uss-zumwalt/.

Anon., n.d. Project 885 Yasen – Multipurpose Nuclear Submarine with Cruise Missiles Severodvinsk (Проект 885 «Ясень» - Многоцелевая Атомная Подводная Лодка с Крылатыми Ракетами «Северодвинск»). Available from army.lv/ru/proekt-885/709/759.

Appolonov, E.M., Kudrin, M.A., Maslich, E.A., Shaposhnikov, V.M., 2002. Fatigue strength estimation of a ship hull with a developed PMC superstructure. In: Proceedings of the Scientific Conference on Strength of Ships Devoted to the Memory of Professor P.F. Papkovich, SPb, November 25–26. (Оценка Усталостной Прочности Судового Корпуса с Развитой Надстройкой из Полимерного Композиционного Материала).

Appolonov, E.M., Kudrin, M.A., Maslich, E.A., Nikolaev, L.S., Fedonyuk, N.N., 2006. A development of design and strength research of a joint of composite superstructure with metal hull. Strength & Design of Surface Ships of PMC, Works of KSRC 27 (311), 71–84. (Разработка Конструкции и Исследование Прочности Узла Соединения Надстройки из Полимерных Композиционных Материалов с Металлическим Корпусом).

Appolonov, E.M., Kudrin, M.A., Maslich, E.A., Fedonyuk, N.N., Shaposhnikov, V.M., 2011. Increase of fatigue strength and watertightness of long superstructures of naval vessels with application of expandable composite joints. In: The 6th International

Conference "Navy & Shipbuilding Nowadays", NSN'2011 Strength Problems of Surface Vessels & Submarines, July 1, Russia, SPb. (Повышение Усталостной Прочности и Герметичности Длинных Надстроек Надводных Кораблей на Основе Применения Конструкций Расширительных Соединений из Полимерных Композиционных Материалов).

Arkhipov, A.V., Bulkin, V.A., Lazarev, A.M., Ogloblin, Y.F., 2006. Application of PMC for naval surface vessels (effectiveness and perspectives). Strength & Design of Surface Ships of PMC, Works of KSRC 27 (311), 6–22. (Применение Полимерных Композиционных Материалов в Надводном Кораблестроении - Эффективность и Перспективы).

Atalex, 2012. Forum militaryrussia.ru, Domestic military equipment (after 1945), Februray. (Буксировка надстройки очередного корвета типа "Стерегущий" к месту достройки). Available from http://militaryrussia.ru/forum/viewtopic.php?p=75920.

Bmpd, 2012. Superstructure for Corvette *Sovershenniy* is Delivered to Komsomolsk-on-Amur, Blog, Center of Strategies and Technologies Analysis, October 12 (Надстройка для Корвета "Совершенный" Доставлена в Комсомольск-на-Амуре). Available from, http://bmpd.livejournal.com/353558.html?thread=7456022.

Bulkin, V.A., Kozlov, C.D., Lebedeva, G.N., Ryzhkin, A.E., Fedonyuk, N.N., 2006. Design & strength of superstructure of PMC. Works of KSRC 1 (27(311)), 23–43, Russia, SPb (Конструкция и Прочность Надстройки из Полимерных Композиционных Материалов).

Bulkin, V.A., Golubev, K.G., Fedonyuk, N.N., 2011. Experience in operating the superstructure made of polymer composite materials on «Corvette» class ships. Works of KSRC 1 (58(342)), 127–136, Russia, SPb (Анализ Строительства и Опыта Эксплуатации Надстройки из Полимерных Композиционных Материалов на Корабле Класса «Корвет»).

Bulkin, V.A., Fedonyuk, N.N., Shlyahtenko, A.V., 2013. Application of perspective composite materials in surface shipbuilding. Morskoy Vestnik 1 (45), 7–8. (Применеие Перспективеых Композиционных Материалов в Надводном Судостроении).

Chatterton, P.A., Paquette, R.G., 1994. The sea shadow. May, Naval Engineers Journal 1–16. Available from, http://www.hnsa.org/seashadow/doc/seashadowASNE.pdf.

DIMMI, 2011. Pr. 20380 – Steregushchy, Military Russia. Available from http://militaryrussia.ru/blog/topic-450.html.

DIMMI, 2012. Pr. 1906 Search-6 – Submersible (пр. 1906 Поиск-6). Available from http://militaryrussia.ru/blog/topic-552.html.

Dronov, B.F., 2002. Trends in development of submarine architecture. Military Technical Almanac "Typhoon", 2 (42) (Тенденции Развития Архитектуры Подводных Лодок). Available from http://flot.com/science/hull/subsarchitecturetrends/.

FAS, 2011. LPD-17 San Antonio Class (formerly LX Class). Available from http://www.fas.org/programs/ssp/man/uswpns/navy/amphibious/lpd17.html.

Fedonyuk, N.N., 2006. Determination of effective characteristics of structural orthotropic middle layer of three-layer panels of the superstructure. Works of KSRC 27 (311), 44–77 (Определение Эффективных Характеристик Конструктивно-Ортотропного Среднего Слоя Трехслойных Панелей Надстройки и Выбор Его Рациональной Структуры).

Gardiner, G., 2012. Composite solutions: cutting cost of nuclear-powered subs. Composites Technology, February 1. Available from www.compositesworld.com/articles/composite-solutions-cutting-cost-of-nuclear-powered-subs.

Greene, E., 1999. Marine Composites, second ed. Eric Greene Associates, Inc., Annapolis, MD. 377 pp. Available from http://ericgreeneassociates.com/images/MARINE_COMPOSITES.pdf.

Greene, E., 2006. The history of submarine composites. May, Composites Manufacturing, 20–26.

Hackett, J.P., 2011. Composites road to the fleet – a collaborative success story. Special Report 306: Naval Engineering in the 21st Century. The Science and Technology Foundation for Future Naval Fleets, Northrop Grumman Shipbuilding – Gulf Coast, 36 pp. Available from http://onlinepubs.trb.org/onlinepubs/nec/61810Hackett.pdf.

Hooper, C., 2010. Virginia class: when does hull coating separation endanger the boat? Next Navy. Available from nextnavy.com/virginia-class-when-does-hull-coating-separation-endanger-the-boat/.

Korablev, D., 2010. Project 22350. Stealth Machinery. Reality & Prospects (Проект 22350). Available from http://paralay.net/22350.html.

Kudrin, M.A., Maslich, E.A., Shaposhnikov, V.M., 2011. Assessment of stress-strain state of long composite superstructures. Works of KSRC 1 (58(342)), 55–58, Russia, SPb, KSRC [in Russian].

Lackey, E., Hutchcraft, E., Vaughan, J., Averill, R., 2006. Zapped electromagnetic radiation and polymeric composites. Composites Manufacturing, May, 7 pp.

Le Lan, J.Y., Livory, P., Parneix, P., 1992. Steel/composite bonding principle used in the connection of composite superstructures to a metal hull. In: Proceedings of SAND-WICH2, Gainesville, Florida, USA.

LeGault, M.R., 2010. DDG-1000 Zumwalt: Stealth Warship – U.S. navy navigates radar transparency, cost and weight challenges with composite superstructure design. Composites Technology. February. Available from http://www.compositesworld.com/articles/ddg-1000-zumwalt-stealth-warship.

Levy, G., 2013. Zumwalt-Class Destroyer Visited by Sec. Hagel, UPI BLOG, November 25. Available from http://www.upi.com/blog/2013/11/25/Zumwalt-class-destroyer-visited-by-Sec-Hagel/4321385389989/.

Lundquist, E., 2012. US Navy: DDG 1000's composite deckhouse milestone. Maritime Reporter & Engineering News 26–28, January.

Mikhailov, A., 2012. Russian Navy receives carbon fiber Stealth ship. Izvestia, Russia, October 9.

Morylyak, A.V., 2009. Stealth-Technology in Shipbuilding, November 30 (Стелс-технологии в Судостроении). Available from www.propulsionplant.ru/content/21/stati/stati-studentov/stels-tehnologii-v-sudostroenii.html.

Mount, M., 2005. Official: U.S. Submarine Hit Undersea Mountain, U.S., CNN.com, January 11. Available from http://www.cnn.com/2005/US/01/10/nuclear.submarine.update/.

Mouritz, A.P., Gellert, E., Burchill, P., Challis, K., 2001. Review of advanced composite structures for naval ships and submarines. Composite Structures 53 (1), 21–42.

Nye, J., 2012. Declassified $170 million Cold War Stealth boat called the Sea Shadow is snapped up for $2.5 million. . . but you can't take it for a spin round the bay. Mail Online, August 9. Available from, http://www.dailymail.co.uk/news/article-2185831/Declassified-170million-Cold-War-Stealth-boat-snapped-2-5million-condition-scrap-parts.html.

CHAPTER 3

Material-Transition Structures

3.1 PREREQUISITES OF RATIONAL DESIGN

A robust weight and cost efficient material-transition structure and a hybrid joint of composite and metal structural components, relevant to a hybrid ship hull application in particular, represents a critical attribute of the hybrid structural system. As ascertained earlier, a hybrid hull is capable of facilitating significant improvement in key performance parameters of a primarily metal warship, including weight saving, increased speed and/or range, and superior signature control. Along with functional, operational, and combat advantages, this is allied with a considerable cost saving in construction, operation, and maintenance, and hence in overall ownership of the ship.

In order to provide these technical and cost benefits, a material-transition structure must satisfy a number of performance requirements, essentially replicating the requirements for regular mono-material structures, either metal or composite, and partly exceeding them. Primary requirements are:

- provision of structural robustness sufficient to withstand a diversity of force and environmental exposures relevant to a ship's normal and combat operations;
- structural integrity with the entire structural system;
- long-term load-bearing capability consistent with or exceeding the hull's assigned length of service, ensuring structural superiority of the material-transition and prevention of its failure;
- watertightness and corrosion resistance;
- standard maintainability and reparability;
- manufacturing cost commensurate with that of mono-material structures.

Similarly to composite-to-composite joints, hybrid joints are typically allied with discontinued fiber reinforcement within the joint; stress concentration attributable to the uneven geometry of a joint structure and/or abrupt alteration of properties of utilized materials; and decreased material performance as a result of secondary (post-cured) bonding, if any.

These adverse traits substantially complicate satisfaction of the imposed requirements for structural joint robustness and efficiency. Also, due to the distinct mechanical and thermal properties of its dissimilar components, a hybrid joint, along with a direct operational force–ambient exposure, experiences internal thermomechanical interaction between those components. This may considerably alter the joint's stress state and affect its long-term structural performance. It is particularly meaningful as the coefficients of thermal expansion of the joint's components are substantially different. In this case, the stress being induced in the joint may notably deviate from its normal state under standard ambient conditions.

Moreover, dissimilarity of the fatigue characteristics typically intrinsic to metal and composite parts may also notably affect long-term performance of a hybrid joint. In particular, this could result in possible migration of critically stressed areas responsible for onset of a fatigue failure during joint service.

On the other hand, an ability to pass an applied operational load from a weakened part to another that stays intact under this loading exposure might be rationally utilized by providing some extra safety margin useful in preventing a sudden failure of the joint.

To meet the imposed requirements of structural performance and alleviate the influence of specific adverse traits intrinsic to a hybrid joint, assorted design and analysis measures may and should be employed.

> To meet the demanding requirements for structural performance and to alleviate the influence of specific adverse traits intrinsic to a hybrid joint, assorted design and analysis measures may and should be employed.

Increasing the bonding area of the composite–metal interface and transverse strengthening thereof, e.g., via incorporation of mechanical fastening, are two conventional ways to combat the possible negative effects relevant to structural performance of hybrid joints.

Stress/strain state analysis needs be carried out, taking into account not only the diverse mechanical properties of utilized dissimilar materials but also their thermomechanical interaction due to temperature alteration during hybrid structure operation. Common math models may need to be revised to address distinct structural behavior of the hybrids under long-term force–ambient operational exposure. The relevant design rationales and required advancement of analytical models are discussed below.

3.2 BENCHMARKING OF EXISTING HYBRID JOINING TECHNOLOGIES

The shipbuilding industry has been dealing with design and manufacturing of heavy-duty hybrid joints for decades. These joints consist of either an assembly of hybrid hull structures, such as those presented in Chapter 2, or installation of metallic hardware on the composite structures of full composite vessels. A broad assortment of conceivable hybrid joining options have been conceptualized to date, many of which are systematically dealt with in Messler (2004).

Despite the great existing multiplicity, just a few types of hybrid joints are suitable for and capable of satisfying the high standards to be met regarding joint producibility, operability, and load-bearing capability with respect to a ship hull's long-term heavy-duty application.

Primarily the methodologies involved are either ordinary bolting or combined bonding-bolting techniques. The conventional plain adhesive bond may also be considered for a ship's application. However, because it is prone to sudden failure, subject to low strain-to-failure ratios, and deficient in its ability to absorb impact energy, this option has limited application and is primarily used for light-loaded and/or auxiliary structures.

The existing bolted and bonded-bolted joints, on the other hand, are substantially more weighty, labor intense, and expensive than plain bonds, due to the need for arduous hole drilling and bolt-nut coupling operations.

3.2.1 Plain Adhesive Bonding

Plain adhesive bonding represents the simplest hybrid joining option, the principal processing steps of which consist of surface preparation of the two metal and composite adherends; placement of an adhesive between those adherends; and solidification of the applied adhesive to produce the bonding film. The type of joint involved has a long history of development and application for composite and hybrid structures, including auxiliary ship structures.

Most of the early work on adhesive joining of composites was done in the 1970s and early 1980s. Since then, a number of research studies aimed at design advancement and improvement of material processing techniques have been conducted. Upgraded analytical models and numerical methods have been developed and widely used.

Matthews et al. (1982) and Banea and Da Silva (2009) provide an extensive overview of the existing design, processing, and analytical

approaches for stress–strain analysis and serviceability evaluation of the adhesively bonded joints. A comprehensive BONDSHIP R&D project (Weitzenböck and McGeorge, 2005), specifically targeted to provide a systematic methodological basis for implementation of adhesive bonding for ship structures, has been recently fulfilled by thirteen partners from seven European nations, including *Vosper Thornycroft*, United Kingdom— the world's leading shipyard specializing in the construction of composite ships. *Det Norske Veritas* (DNV) coordinated the project, the principal goal of which was to make European shipyards more competitive by achieving considerable cost savings in the production of high-speed craft and passenger ships. The primary focus of the effort was on aluminum-aluminum/steel and aluminum-composite bonding joints of lightweight materials for cost-effective ship production. The guidelines (Weitzenböck and McGeorge, 2005) summarize the results of the BONDSHIP project and the steps necessary to design, build, inspect, and repair bonded joints in ships.

Several factors govern structural performance of a plain adhesive bond undergoing assorted force-ambient operational exposures. The principal ones are surface preparation of the adherends, the shape of the joint components, the properties of utilized materials, and the bond/PMC processing.

The surface preparation of the metal adherend prior to its consolidation with the composite typically comprises grid/sand blasting, abrasion/solvent cleaning, and priming. When properly executed, these procedures facilitate a sturdy adhesive bond between the metal and composite.

> Properly executed surface preparation of the metal adherend prior to its consolidation with the composite is the main factor enabling the required sturdiness of the adhesive bond between the metal and composite.

This is accompanied by an increase in the surface energy of the adherends and formation of chemical bonds—mainly covalent (but some ionic and static attractive bonds may also be present)—between the adherend surface atoms and the compounds constituting the adhesive (Baldan, 2004).

Three conventional configurations of adhesively bonded hybrid joints, single-lap, composite double-lap, and metal double-lap, embody the commonly used joining options suitable for ship hull application. Figure 3.1 delineates these three traditional design options.

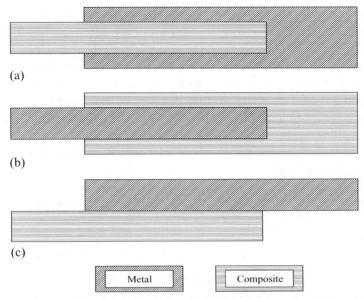

Figure 3.1 Three principal adhesively bonded hybrid joint configurations. (a) metal double-lap, (b) composite double-lap, and (c) single-lap. *Adapted from Dance and Kellar (2010).*

Note that the sketches do not reflect the actual shape of the adherends. In reality, the latter are usually tapered to minimize material consumption and have geometric irregularities smoothed down, thereby reducing stress concentration on the joint structure and providing a streamlined hull contour.

As is common for a structural design, each option has certain pros and cons. Specifically, the metal double-lap (Figure 3.1a) is superior to others in structural performance with regard to transverse bending and lateral impact, both of which are typical for prevailing load cases inherent to ship operation. The relative complexity of both manufacturing procedure and assembly for the base metal hull represents the downside of this joint configuration.

Manufacturability and assembly of the composite double-lap joint (Figure 3.1b) are both friendlier than for the metal double-lap, but it is notably inferior with respect to load-bearing capability under transverse bending and lateral impact.

The single-lap joint (Figure 3.1c), because of the substantially smaller bonding area and the asymmetry of the load pass, is substantially weaker than either double-lap option of the same size, although joint production and assembly are much simpler than for those options.

The stress distribution along the material-transition region of an adhesively bonded joint is extremely uneven with the maximum stress at the ends of the overlap and much lower at the middle. Volkersen's (1938) classic differential solution, based on an assumption that the adhesive bonding film deforms only in shear but the adherends can deform in tension, largely reflects the typical character of stress distribution along the bond line.

With regard to a symmetrical double-lap joint configuration, the shear stress within the bonding film in terms of a non-dimensional stress $\tilde{\tau}_a$, referred to the averaged adhesive shear stress (F/b_a), is expressed in the Volkersen math model as

$$\tilde{\tau}_a = \frac{\omega}{2}\left(\frac{\cosh(\omega\tilde{x})}{\sinh(\omega/2)} + \frac{2t_p - t}{2t_p + t}\frac{\sinh(\omega\tilde{x})}{\cosh(\omega/2)}\right), \quad -0.5 \le \tilde{x} \le 0.5 \qquad (3.1)$$

where

$$\omega = b_a\sqrt{\left(1 + 2\frac{t_p}{t}\right)\frac{bG_a}{Et_p t_a}} \qquad (3.2)$$

Here b_a is the adhesive bond extent; t, t_p, and t_a are thickness values of the inner (overlapped) and outer adherends and the adhesive film, respectively; E is Young's modulus of the inner adherend; G_a is the shear modulus of the adhesive film; and $\tilde{x}$ is the relative distance from the center of the overlap extent.

Figure 3.2 illustrates the generalized shear stress distribution along the bond line of a metal double-lap joint and signifies the peak stress at the ends of the material transition region which defines the joint's load-bearing capability.

It should be realized that, while notionally correct, the Volkersen simplified solution does not reflect an actual joint design or specific loading conditions which may notably affect the stress state within a real joint. The actual adherend shape, an apposite alteration of material properties within the joint, and the formation of a shaped spew of excess of the adhesive squeezed out of the lap region can sizably downgrade the stress disproportion within the joint, thereby improving its load-bearing capability.

Extensive studies have been carried out to advance the math models to accurately assess the stress state of the adhesively bonded joints. The references given here (Goland and Reissner, 1944; Renton and Vinson, 1975; Allman, 1977; Adams et al., 1997; Weitzenböck and McGeorge, 2005; Gustafson et al., 2006) represent a few of the many works dedicated to analytical

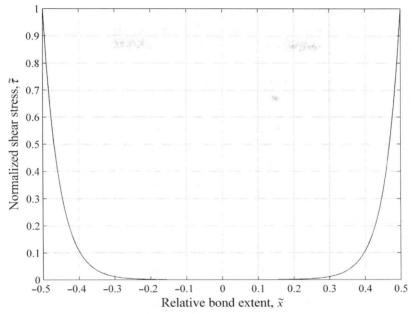

Figure 3.2 Shear stress distribution along the bond-line.

characterization of the stress state within an adhesively bonded joint. This is done taking into consideration the joint's actual geometry, which is among the most influential factors on stress distribution along the bond line and ultimately on the joint's load-bearing capability. The results to date clearly indicate that a properly selected configuration may help to notably reduce the peak stress at the joint ends. Some stresses, such as transverse tension and interlaminar shear, causing peel and cleavage—typical failure modes for a hybrid joint—may and should be minimized, whereas others, such as bearing stress, may be safely maximized.

> Properly selected configuration may help to notably reduce the peak stress at the joint's ends; some stresses, such as transverse tension and interlaminar shear, causing peel and cleavage, may and should be minimized, whereas others, such as bearing stress, may be safely maximized.

Two principal material processing options for bond formation are available. One is executed simultaneously with the processing of the non-cured PMC adherend and its consolidation with the metal counterpart. The other

involves a secondary bonding of the post-cured PMC adherend with the metal. The first, co-curing, is a superior option with regard to joint structural performance, associated with the sound integrity of heterogeneous components and a comparatively low scattering range of the attained strength parameters. The second option is usually much more assembly friendly than the first processing option, while being inferior with respect to joint structural performance.

The primary cause of this distinction is a fit-up problem between large metal and composite structure parts allied with different degrees of thickness variation of the bonding film. In the first processing option, the fit-up problem is resolved naturally, as the non-cured PMC being processed perfectly matches the actual shape of the metal adherend. The secondary bonding on the other hand induces a considerable thickness variation of the adhesive film, which notably downgrades the joint's performance.

The influence of the adhesive film's thickness variation manifests primarily in two ways. One is direct, as reflected in Equations (3.1)–(3.2). Another shows itself in the effective mechanical properties of the bond film, which are impaired with the increase in thickness. These adverse effects are aggravated by another detrimental trait inherent to plain bonds. It is a tendency to persistent propagation of a debonding crack due to the lack of stoppers, such as fasteners being used for combined joints.

Altogether, the noted peculiarities bring considerable uncertainty to quantification of the load-bearing capability of a plain adhesive bond despite the accuracy of idealized math models being employed to characterize the stress distribution within the joint. This in turn requires an increase in safety margins, which ultimately results in lowered structural efficiency of both the joint itself and the entire hybrid structure.

Ultimately, the noted adverse traits preclude the use of plain adhesively bonded joints (especially the secondary bonding option) for a ship's heavy-duty applications in spite of the great advantages in assembly operations and potential weight and cost savings.

One of the rare examples of plain adhesive bonding use for a naval hybrid hull structure is the GFRP deckhouse and deck structure set on the metal deck of the *La Fayette* class frigate (Le Lan et al., 1992; Greene, 1999).

Boyd et al. (2004), who performed a study on the structural integrity of the adhesive composite double-lap joint based on the *La Fayette* application case, reported experimental data gained from static and fatigue testing of the plain adhesive joint. The joint test specimens consisted of part of a regular sandwich panel of the main deckhouse structure, with 4-mm-thick skins of

GFRP; a tapered part of the sandwich with converging GFRP skins; the 240-mm extent of material transition—the bonding region, where GFRP skins overlapped the metal middle plate; and only metal plate beyond the material transition.

The composition of the GFRP skins comprised vinyl ester resin (Dow Derakane 411–C50) reinforced with the 3×1 twill weave 780 g/m^2 E-glass woven roving (Chomarat 800S4). The sandwich's core consisted of 150 kg/m^3 balsa wood (Baltek AAL600-10 Contourcore). The metal plate was made of the 6–mm–thick mild steel (D55).

As reported by Boyd et al. (2004), the 100–mm–wide joint test specimens underwent compression loading to failure, which resulted in an ultimate static compression load of 108 kN, or a linear ultimate load 1080 kN/m of the joint extent (along the specimen's width).

While the reported experimental data provide valuable information on load–bearing capability of this type of hybrid joining under compression loading, the in–plane tension loading mode seems to be a critical load case relevant to a joint behavior during warship operation. In part, this is due to the in–plane tension loading being associated with the transverse tension of the composite overlaps and their tearing off the metal plate, which, along with laminar shear, cause debonding at the joint's interface and/or delamination of the composite part thereof governing a hybrid joint's performance. In contrast, the composite overlaps do not experience transverse tension under in–plane compression. In–plane compression induces transverse bearing within the material transition region, which is the most tolerable stress component for laminar PMCs.

To evaluate the robustness of the plain bond under in–plane tensile loading, a metal double–lap hybrid joint was tested by applying a tension force capable of separating the joint's components (Shkolnikov et al., 2009). The joint specimens failed under a static force of 1944 ± 35 kN/m of the joint extent. The ultimate state of the tested joint was associated with a mixed adhesive-cohesive failure mode, or, in other words, a combination of a failure at the adhesive film–adherend interface and a failure of the adhesive film itself. This testifies to the necessity for properly prepared contact surfaces of the adherends prior to their bonding. A more detailed description of the tests performed, along with a summary of the experimental data, will be presented in Chapter 4.

Wright et al. (2000) investigated the structural behavior of bonded steel–to–composite joints with regard to their application for auxiliary composite structural components in large steel ships. The reported experimental data, as well as the outcome of computer simulations, proved the feasibility of the

bonded joints and provided useful input for the application of plain adhesive joints to light double-skinned composite sandwich panels in steel ships.

3.2.2 Bolted Joints

The conventional type of fastened (bolted) joints are usually preferred for a heavy-duty application because of superior load-bearing capability and good predictability and controllability of the joint's structural performance. In addition, the methodology for the determination of design parameters of bolted joints is very well established and minimizes possible uncertainties in the serviceability evaluation of the hybrid joints.

Ordinary bolting represents the basic joining option being employed for the mounting of composite structural components on primarily metal hulls for both major categories of warships, surface vessels and submarines. Photographs of the bow dome and sail cusp prepared for installation on the metal hull of a *Virginia* class sub and those for mounting the metal-skirted composite topside structures of the *Zumwalt* class destroyer, presented in references (see Anon, 2012, 2013; Lundquist, 2012; Tortorano, 2011), well illustrate the US Navy's inclination to use bolted joints for heavy-duty applications.

Meanwhile, the bolted joints are also far from an ideal joining option for a ship's hybrid hull application. The holes drilled within a composite part to insert the bolts are allied with multiple cuttings of the fiber material—the primary load-bearing element of FRPs. This substantially impairs the composite part, not just because of interrupting fiber continuity but also by creating multiple origins for stress concentration, which is especially significant for the relatively big holes required for the heavy-duty applications. To compensate for this, both the thickness of the composite part and the sizing of the bolts have to be substantially increased. As a result, the weight of bolted joints increases, and their structural efficiency declines.

It should also be noted that the achievable strength of a bolted joint under quasi-static loading is substantially higher than that of a plain adhesive bond; whereas the long-term fatigue strength related to normal ship operation loading of these two options is comparable.

> Although achievable ultimate strength of bolted joints under quasi-static loading is substantially higher than that of plain adhesive bond, the long-term fatigue strength related to normal ship operation loading is comparable for these two joining options.

As a matter of fact, the fatigue performance is typically the main criterion for selection of a preferred hybrid joint option, as a specific military–origin impact/shock loading is not the dominant operational load case.

One more significant negative factor applying to bolted joints is problematic water sealing and associated intense (crevice) corrosion of the bolts and adjacent metal details. This also hampers utilization of ordinary bolted joints for naval platforms.

Ultimately, an application of the disproportionately heavy, bulky, and costly composite-metal assemblage utilizing a bolted joint may result in noticeable cutback of the anticipated functional advantages of PMC utilization for primarily metal vessels. Per Brown's (2004) estimation, the weight impact of the bolted joints employed for *Zumwalt* class destroyers was quite significant, at 162 kg/m of the composite-metal seam extent.

The compromised weight benefits of the PMC application, excessive labor operations, related high manufacturing cost, and problematic water-sealing associated with mere bolting are probably primary reasons for the US Navy's decision to reject the potentially advantageous composite option of the deckhouse structure for the *Lyndon B. Johnson* (DDG 1002), the third and last unit of the stealthy *Zumwalt* class of destroyer, switching to a less costly steel deckhouse alternative (Cavas, 2013a,b). Conceivably, this changeover will notably downgrade the stealth performance of that vessel.

3.2.3 Bonded-Bolted/Fastened Joints

Along with plain adhesive bonding and mere bolted joints, one more basic joining technique is used to mount composite structures on a primarily metal ship hull. This is a combined bonding-fastening that implies a substantial load-sharing between the two principal load-bearing counterparts, adhesive bond and mechanical fasteners.

Bolts embody the most common type of fasteners applicable for heavy-duty joints, while rivets and wood screws may alternatively be used to provide the requisite transverse reinforcement of the joint body pertinent to moderately loaded structures.

The combined joining technique can potentially greatly decrease the shortcomings inherent to separate applications of an ordinary joining method, plain adhesive bonding or mere bolting. More than that, a synergetic interaction of the two can considerably increase the load-bearing capability and structural efficiency of a combined joint. Ideally, this occurs as the adhesive bond bears most of the applied load so that the fasteners do not

experience substantial loading but rather prevent premature debonding of the joint by keeping the adhesive bond integral.

Apparently, while fasteners are intact the bonding film is sound and capable of bearing a major portion of the load. Concurrently, while the adhesive film is unharmed, at least some fasteners should stay load-bearing. A metal fastener will continue to bear its partial load even upon achieving a yielding state, allowing redistribution of its load share between adjacent fasteners that are relatively underloaded. The presence of the adhesive bond enhances this interaction of fasteners, leveling their stressing and increasing thereby the load-bearing capability of the entire joint.

To implement such a scenario, the fasteners do not need to be as large and strong as those in the ordinary bolted joints, independently bearing the entire applied load, but rather may be relatively tiny, sufficient just to maintain the bond's integrity. The term "chicken rivets," often used for fasteners of a combined joint, well reflects this distinct role of mechanical fastening within a combined bonded–fastened joint.

Thanks to the drastic size reduction, stress concentration at the bolt holes significantly alleviates the need for increased thickness of the coupled adherends within material transition (the joint structure). It also favorably manifests better fatigue performance of combined joints when compared to ordinary bolting.

Besides, the combined bonded-bolted joints possess watertight integrity and provide corrosion protection in a natural way by filling up all gaps and cavities with the adhesive, including areas inaccessible for inspection.

Conjointly, the design advantages intrinsic to combined bonded-bolted joints facilitate increased serviceability and structural efficiency of these joints along with notable reduction of the weight, labor intensity, manufacturing cost, and maintenance expenses when compared to bolted-only joints.

> Conjointly, the design advantages intrinsic to combined bonded-bolted joints facilitate increased serviceability and structural efficiency of these joints while notably reducing weight, labor intensity, manufacturing cost, and maintenance expenses when compared to bolted-only joints.

The combined joints also have a long record of successful applications for naval structures, starting in the 1960s. This particular type of hybrid joint was used to mount external GFRP panels on the metal hulls of several classes of

Soviet subs, specifically: the Project 651 *Juliet*, the Project 627A *Kit/November*, and the Project 865 *Piranha/Losos* class subs introduced in Section 2.2.

Presumably, a conceptually similar bonded-bolted joint configuration is presented in a patent (Nikolaev et al., 2011) that is being utilized to mount composite deckhouse and hangar structures on the metal deck of the *Steregushchy/Tigr* class corvettes. Photos presented in reference Bmpd (2012) of the composite deckhouse of the *Sovershenny*, the lead ship of the second batch of *Steregushchy* class corvettes, built in Komsomolsk-on-Amur, testify well for the much more slender body of the material-transition structure of her PMC deckhouse compared to that of similar embodiments of the PMC topside structures of *Zumwalt* class vessels, shown by Tortorano (2011) and Lundquist (2012).

While the advantages of the combined bonding-fastening are recognized, it is not feasible to execute a routine bolted joint design and material processing. To provide the sought increase of structural efficiency allied with load-sharing between the joining counterparts (adhesive bond and mechanical fasteners) the joint body, including metal skirt and bolting, is co-processed with the base composite structure.

Further, math models for determination of the stress-strain state and strength reconciliation of the combined joints need to be properly tweaked to reflect the pursued load-sharing. Otherwise, sticking with the conventional conservative design that demands that at least one load-bearing counterpart, either the adhesive bond or fasteners, is capable of bearing the entire applied load is the only option, thereby ensuring the joint's structural robustness. Obviously, such structural redundancy is inconsistent with the pursued load-sharing and unavoidably affects structural efficiency of the hybrid joint, inflicting the penalties of extra weight and high construction cost, ultimately diminishing the benefits of the PMC application.

Structural redundancy regarding strength reconciliation of combined bonded-bolted joints is inconsistent with the pursued load-sharing and unavoidably affects the structural efficiency of the hybrid joint, leading to penalties for extra weight and a high construction cost ultimately diminishing the benefits of the PMC application.

It should be noted that despite the great performance superiority of the combined joint over ordinary joints with separate utilization of either plain adhesive bonding or mere bolting, it is nevertheless associated with massive

drilling and bolt–nut coupling, which intensify the labor operations and raise the cost of manufacturing when compared to those of the plain adhesive bond.

3.3 ADVANTAGEOUS JOINING OPTIONS

Due to imperfection of the existing technologies, innovative hybrid joining options capable of suppressing the faced deficiencies and promising superior structural performance along with moderated cost are being continually pursued. One of the recent attainments is an advanced adhesive joint being promoted by the US Navy ONR's *Navy Joining Center* (NJC), operated by *Edison Welding Institute* (EWI) in Columbus, Ohio (Simler and Brown, 2003). Another is Comeld, an innovative joining technology pioneered by *The Welding Institute* (TWI), Cambridge, United Kingdom (Dance and Kellar, 2004; Dance and Kellar, 2010; Smith, 2004). More specifically, it is a "Comeld–2" version of the Comeld, which is conceptualized for naval vessel applications by *Concurrent Technologies Corporation* (*CTC*) of Johnstown, Pennsylvania (Shkolnikov et al., 2009) and further developed by *Beltran, Inc.*, Brooklyn, New York (Shkolnikov and Khodorkovsky, 2011).

3.3.1 NJC's Novel Adhesive Bonding Joint

As Simler and Brown (2003) assert, the NJC's novel hybrid joining concept embodied a modified conventional adhesive bonding for application in large composite topside ship structures. Specifically, the reported effort was to develop a superior and cost effective method for adhesive bonding of a composite deckhouse of the DDG 1000 class destroyer to its steel hull, sufficient to replace the conventional bolting technique, which met the tough service requirements but was excessively expensive.

The new joint design in Simler and Brown's (2003) description involves a PMC balsa–wood–cored sandwich structure that fits into a flat-bottomed steel shoe with slightly tapered sides in the lower part of the engagement. A paste adhesive is used to form the bond between the steel shoe and the outside faces of the composite material. This design essentially replicates the traditional adhesive joint referred to in the patent of Grose et al., 2004, with the prior art being adapted to coupling of a metal base with a sandwich PMC structure.

Brown (2004) asserts that the NJC's joint design showed that any failure of the joint was due to stresses occurring within the composite material itself and not in the adhesive joint. The design was to maintain the joint in

compression mode during all loading occurrences. Fatigue tests were performed to simulate a 35-year service using 1.1 million cycles of load. It was ascertained that the joint did not suffer any fatigue and had the same failure load values after the fatigue cycles as it had before the test.

Specifically, the reported results of testing performed by EWI show that the newly-promoted joint design could withstand combined static in-plane and seaway loads of at least $F_U = 1500$ kN/m at a room temperature of $T = 23\,°C$ and $F_U = 1000$ kN/m at the elevated temperature of $T = 60\,°C$, with no internal corrosion after 60 days of hot-wet environmental exposure and 1500 h of salt fog exposure. This was enough to win the method permanent acceptance as a DNV class joint for bonding composite to metal structures, according to Gardiner (2009).

Meanwhile, it should be noted that the reported combined test load mode was not specified and the given results were probably irrelevant to the solely tensile (pulling) loading critical for structural performance of a butt hybrid joint. Nevertheless, if we assume that the given ultimate strength values pertained to in-plane tensile loading, it was nevertheless substantially lower than that being withstood by either merely bolted or combined bonded-bolted joints typically employed for heavy-duty naval structures. Relevant data, courtesy of the *Naval Surface Warfare Center Carderock Division* (NSWCCD), are presented in Shkolnikov and Khodorkovsky (2011) and reflect results of in-plane tensile testing of a bolted metal-double-lap hybrid joint destined for the same application— to mount the composite deckhouse of the DDG-1000 class destroyer. The given mean ultimate linear force was $F_U = 5180$ kN/m, i.e., 3.45 times that of the NJC's novel joint.

While inferior to the bolted joint standard in load-bearing capability, the NJC's adhesive joint, like the conventional adhesively bonded joint, weighed much less than the bolted joint. As assessed by Brown (2004) with regard to the *Zumwalt*'s deckhouse application, the NJC's adhesive joint weighed 94 kg/m, vs. 162 kg/m relevant to the used bolted joint. Thus, 68 kg/m weight savings might be available, as the NJC's new adhesive joint possessed sufficient robustness corresponding to a naval application. Especially concerning were high sea state operation and underwater shock exposures typical for a naval service. Each of these extreme loadings is able to produce significant pulling forces critical for the joint's performance.

The extensive machining of the metal shoe allied with considerable extra expenses signified another noticeable shortcoming intrinsic to the NJC's joining concept.

Despite all the promise and concerns related to the NJC's adhesive joining option, the shipyards did not buy it and the DDG 1000 composite deckhouse was produced using the conventional bolted joining technique.

Despite this, the NJC's joining technology has found a civilian application in the private sector. It is the *Swift 141* luxury gigayacht being converted from a Dutch 130-m, 3,500-ton displacement S-class frigate. The yacht is furnished with a large, 100-m-long, 14.4-m-wide, and 13.5-m-high composite superstructure, and the NJC's joint was used to mount that superstructure to the steel deck of the *Swift 141* (later renamed *Yas*) yacht—the eleventh largest yacht on the water today (Gardiner, 2009; Liversedge, 2011).

3.3.2 Comeld Hybrid Joining Technology

The innovative Comeld hybrid joining technology combines adhesive bonding with mechanical "pinning" of the composite by upright features protruding from the metal substrate (Smith, 2004). Such a bonded–pinned joint enables substantial increase of the load–bearing capability of the adhesive bond, considerably exceeding structural efficiency of other joining options available to date.

The electron beam (EB) Surfi–Sculpt process (Buxton and Dance, 2005) is used to cause protrusion of the metal substrate with the desired pattern needed to constitute a bonded–pinned joint. Using EB, a material is processed to give a number of different types of surface topography as defined by patent (Dance and Kellar, 2010). Highly automated, easily computer-controlled and low labor–intensive EB can reshape materials precisely, "growing" projections that rise from the surface of the material, just marginally affecting the manufacturing cost of an adhesively bonded hybrid joint.

Photographs presented in references (Buxton and Dance, 2005; Dance and Kellar, 2010) exemplify protrusion patterns suitable for a Comeld joint application.

Contrary to potentially competing technologies, such as direct metal deposition, Surfi–Sculpt is not an additive process and therefore does not require complicated powder or wire-feed delivery systems. Movement of the material is thought to be dependent upon both vapor pressure and surface tension forces. The process mechanisms are fundamentally comparable to keyhole welding, whereby a highly focused power beam produces a vapor cavity in a substrate that is translated through the work piece, according to the motion of the beam. Traversing the beam across the work piece creates a

trailing melt pool. The combination of surface tension variation along the weld pool and vapor pressure from the cavity causes a displacement of material opposite to the beam's direction of travel, allowing formation of a crater at the end of the weld and a corresponding bump at the weld initiation point. The Surfi-Sculpt process employs in–vacuum EB, which can be highly focused and quickly deflected using computer-controlled electromagnetic coils.

In the original Comeld configuration (Buxton and Dance, 2005; Smith, 2004), EB-protruded metal is embraced by the composite overlaps forming a bonded–pinned composite double-lap joint. Experimental data presented by Buxton and Dance (2005) demonstrate Comeld's capability of increasing the load-bearing of an adhesive bond under in–plane tension up to 72%, while also improving energy absorption under dynamic loading when compared to an outwardly similar plain bonded joint.

While conceptually advantageous, the Comeld original configuration is not optimum for transverse bending, a primary load case for ship hull application. To overcome this discrepancy, the original Comeld was reconfigured to the metal double-lap joint option. This alteration was conceptualized by *CTC* in cooperation with *TWI* under an ONR MANTEC contract. Shkolnikov et al. (2009) report that Comeld was reconfigured so as to embrace the composite part by metal lap plates forming a bonded-pinned metal-composite assemblage that is stiffer and stronger than its original composite double-lap analogue.

The result provided structural advantages with regard to all major exposures typical for a naval ship operation, including two critical load cases, transverse bending and direct transverse impact. In addition there was substantial improvement of the joint's damage resistance, with metal overlaps protecting the embraced composite from a direct impact.

Figure 3.3 replicates a generalized view of the reconfigured Comeld structure, referred to henceforth as "Comeld-2" to distinguish it from the original option and avoid any related confusion.

Both target traits of Comeld-2 being pursued—feasibility of the relevant manufacturing technology and noticeable superiority of Comeld-2's load-bearing capability when compared to plain adhesive bonding—had been attained as a result of *CTC*'s conceptualization efforts (Shkolnikov et al., 2009). Specifically, ultimate linear force $F_U = 2850 \pm 190$ kN/m under in-plane tension was gained regarding an initially selected not–yet–optimized Comeld-2 configuration, whereas the outwardly identical bonded-only joints used as baseline withstood the ultimate force $F_U = 1940 \pm 40$ kN/m, corresponding to a 48% structural advantage of the novel Comeld-2 joint.

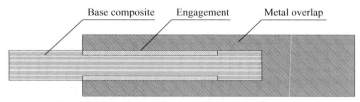

Figure 3.3 Modified bonded-pinned (Comeld-2) configuration. *Adapted from Shkolnikov (2013).*

Two employed material processing techniques, Surfi–Sculpt and VIP, being adapted to construct a large hybrid hull structure, would allow for elimination of the labor–intensive hole drilling and bolt–nut coupling related to the state-of-the-art bolted and bonded-bolted joints. Ultimately, this drastically reduces hand operations and enhances repeatability and reliability of joint manufacturing when compared to existing bolted and bonded-bolted joint manufacturing.

> Two employed material processing techniques, Surfi-Sculpt and VIP, being adapted to construction of a large hybrid hull structure with utilization of Comeld-2, would allow for elimination of the labor-intensive hole drilling and bolt-nut coupling operations associated with state-of-the-art bolted and bonded-bolted joints, while providing great weight savings and superior structural efficiency comparative to those of existing joining options.

Because of the promising outcome of the initial conceptualization study, a further development of the Comeld-2 hybrid joining concept was undertaken. The principal results of that endeavor are presented in Chapter 4.

REFERENCES

Adams, R.D., Comyn, J., Wake, W.C., 1997. Structural Adhesive Joints in Engineering. Springer, 359pp.

Allman, D., 1977. A theory for elastic stresses in adhesive bonded lap joints. Q. J. Mech. Appl. Math. 30, 415–436.

Anon, 2012. Goodrich delivers first composite sail cusp for Virginia class submarine. Reinforced Plasics.com, News, June 12. Available from http://www.reinforcedplastics.com/view/26255/goodrich-delivers-first-composite-sail-cusp-for-virginia-class-submarine/.

Anon, 2013. Marine composite structures, products. UTC Aerospace Systems. Available from http://utcaerospacesystems.com/cap/products/Pages/marine-composite-structures.aspx.

Baldan, A., 2004. Review: adhesively-bonded joints and repairs in metallic alloys, polymers and composite materials: adhesives, adhesion theories and surface pretreatment. J. Mater. Sci. 39, 1–49.

Banea, M.D., Da Silva, L.F.M., 2009. Adhesively bonded joints in composite materials: an overview. Proc. Inst. Mech. Eng., L: J. Mater. Design Appl. 223, 1–18.

Bmpd, 2012. Superstructure for Corvette *Sovershenniy* is Delivered to Komsomolsk-on-Amur, Blog, Center of Strategies and Technologies Analysis, Oct 12 (Надстройка для Корвета "Совершенный" Доставлена в Комсомольск-на-Амуре). Available from http://bmpd.livejournal.com/353558.html?thread=7456022.

Boyd, S.W., Blake, J.I.R., Shenoi, R.A., Kapadia, A., 2004. Integrity of hybrid steel-to-composite joints for marine application. Proc. Inst. Mech. Eng., M: J. Eng. Maritime Environm. 218 (4), 235–246.

Brown, L., 2004. Composite to Steel Joints – Developed for the Next Generation Surface Combatant. Technical Presentation, the ASM International Indianapolis Chapter.

Buxton, A.L., Dance, B.G.I., 2005. Surfi-Sculpt™ – revolutionary surface processing with an electron beam. In: Proceedings of ASM International ISEC Congress, St Paul, MS, 1–3 Aug.

Cavas, C.P., 2013a. Navy-switches-from-composite-steel. Defense News, Aug 2. Available from www.defensenews.com/article/20130802/DEFREG02/308020010/Navy-Switches-from-Composite-Steel.

Cavas, C.P., 2013b. Gulfport-composites-shipyard-close. Defense News, Sep 4. Available from www.defensenews.com/article/20130904/DEFREG02/309040016/Gulfport-Composites-Shipyard-Close?goback=%2Egde_88203_member_271117055#%21.

Dance, B.G.I., Kellar, C., 2004. Workpiece Structure Modification. Number WO 2004/028731 A1. International Patent Publication. Available from https://data.epo.org/publication-server/rest/v1.0/publication-dates/20101103/patents/EP1551590NWB1/document.pdf.

Dance, B.G.I., Kellar, C., 2010. Workpiece Structure Modification. Patent 7667158 B2, USA. Available from http://www.google.co.ck/patents/US7667158.

Gardiner, G., 2009. From frigate to luxury gigayacht. Composites Technology, 15 (5), 32. http://www.compositesworld.com/articles/from-frigate-to-luxury-gigayacht.

Goland, M., Reissner, E., 1944. The stresses in cemented joints. J. Appl. Mech. 66, A17–A27.

Greene, E., 1999. Marine Composites, second ed. Eric Greene Associates, Annapolis, MD. 377pp. Available from http://ericgreeneassociates.com/images/MARINE_COMPOSITES.pdf.

Grose, D.L., Wanthal, S.P., Sweetin, J.L., Mathiesen, C.B., Southmayd, R.A., 2004. Methods of Joining Structures and Joints Formed Thereby. Patent 7,393,488 B2, USA, July 1. Available from https://docs.google.com/viewer?url=patentimages.storage.googleapis.com/pdfs/US7393488.pdf.

Gustafson, P.A., Bizard, A., Waas, A.M., 2006. Dimensionless parameters in symmetric double lap joints: an orthotropic solution for thermomechanical loading. In: AIAA 2006-1959, AIAA/ASME/ASCE/AHS/ASC 47th Structures, Structural Dynamics, and Materials Conference, May 1–4, Newport Rhode island, 17pp.

Le Lan, J.Y., Livory, P., Parneix, P., 1992. Steel/composite bonding principle used in the connection of composite superstructures to a metal hull. In: Proceedings of SANDWICH2, Gainesville, Florida.

Liversedge, B., 2011. 141m Yas Launches at ADM. SuperyachtNews.com, Nov 25. Available from http://www.superyachtnews.com/fleet/17066/yas_launches_at_adm.html.

Lundquist, E., 2012. US Navy: DDG 1000's composite deckhouse milestone. Maritime Reporter Eng. News. 26–28.

Matthews, F.L., Kilty, P.P.F., Godwin, E.W., 1982. A review of the strength of joints in fiber-reinforced plastics 2: Adhesively bonded joints. Composites 13 (1), 29–37.

Messler Jr., R.W., 2004. Joining composite materials and structures: some thought-provoking possibilities. J. Thermoplastic Compos. Mater., 51–75.

Nikolaev, L.S., Bulkin, V.A., Ivanov, I.N., Kacznelson, L.I., Fedonyuk, N.N., 2011. Bonded-bolted Joint of Composite Superstructure with Metal Ship Hull. Patent 2235660, Russia (Клее-болтовое Соединение Надстройки мз Композитного Материала с Металлическим Корпусом Судна). Available from http://www.findpatent.ru/patent/223/2235660.html.

Renton, W., Vinson, J., 1975. The efficient design of adhesive bonded joints. J. Adhesion 7, 175–193.

Shkolnikov, V.M., 2013. Material-Transition Structural Component for Producing of Hybrid Ship Hulls, Ship Hulls Containing the Same, and Method of Manufacturing the Same. Patent 8,430,046 B2, USA.

Shkolnikov, V.M., Khodorkovsky, Y., 2011. To-date advancement of bonded-pinned composite-to-metal joining technology. In: Proceedings of SAMPE-2011 Conference, Long Beach, CA, 23–26 May, 13pp.

Shkolnikov, V.M., Dance, B.G.I., Hostetter, G.J., McNamara, D.K., Pickens, J.R., Turcheck, S.P., Jr., 2009. Advanced hybrid joining technology. OMAE2009-79769. Proceedings of the ASME 28th International Conference on Ocean, Offshore & Arctic Engineering, OMAE2009, Honolulu, Hawaii, May 31–June 5, 8pp.

Simler, J., Brown, L., 2003. 21st century surface combatants require improved composite-to-steel adhesive bonds. AMPTIAC Q. 7 (3), 21–25, Available from http://ammtiac.alionscience.com/pdf/AMPQ7_3ART03.pdf.

Smith, F., 2004. COMELD – an innovation in composite to metal joining. Composites Processing, CPA, Bromsgrove, UK, Apr 23. Available from http://www.twi.co.uk/technical-knowledge/published-papers/comeld-an-innovation-in-composite-to-metal-joining/.

Tortorano, D., 2011. Beyond shipbuilding. Alliance Insight, V(II), Mississippi Gulf Coast Alliance for Economic Development, July. Available from http://www.mscoastaerospace.com/news-publications/files/AI-Q3-2011.pdf.

Volkersen, O., 1938. Die Niektraftverteilung in Zugbeanspruchten mit Konstanten Laschenquerschritten. Luftfahrtforschung 15, 41–68.

Weitzenböck, J.R., McGeorge, D., 2005. BONDSHIP Project Guidelines. Det Norske Veritas, 254pp, ISBN 82-515-0305-1.

Wright, P.N.H., Wu, Y., Gibson, A.G., 2000. Fibre reinforced composite-steel connections for transverse ship bulkheads. Plastics Rubber Compos. 29 (10), 549–557.

CHAPTER 4

Comeld-2 Development and Performance Evaluation

This chapter presents results of the continued development of Comeld-2 hybrid joining technology. In Comeld-2, assorted analytical, design, manufacturing trials, and mechanical-environmental testing, all ultimately targeted to optimization of the material-processing technology and design parameters were implemented. These efforts, sponsored by the ONR (USA), were carried out by *Beltran, Inc.*, Brooklyn, New York, within the framework of a small business technology transfer (STTR) project in cooperation with *CTC* and several other companies and institutions well positioned at relevant industries and market domains to ensure a tight focus on feasible, effective, and navy-suitable technical solutions. Along with the ONR, *CTC*, and *Beltran* these included:

- NSWCCD, Bethesda, Maryland, assigned to provide technical advisory on the investigation subject matter
- *EBTEC Corporation*, Agawam, Massachusetts, a high-tech company specializing in cutting-edge metal-treatment technologies capable of implementing the metal EB Surfi-Sculpt protrusion needed to produce Comeld-2 hybrid joint structures
- *Triton Systems, Inc.* (TSI), Chelmsford, Massachusetts, a material science and engineering company with composite prototyping and manufacturing capabilities suitable for the required fabrication of full-scale composite and hybrid test articles
- *Westmoreland Mechanical Testing & Research, Inc.* (WMT&R), Youngstown, Pennsylvania, a materials research and testing house, assigned to execute the mechanical-ambient test program
- The US *Naval Academy* (USNA), Annapolis, Maryland, entrusted with carrying out independent comparative static and impact testing of Comeld-2 specimens and other hybrid joining options.

The imparted data ensuing from the *Beltran*'s endeavor comprise results of computer simulations, selection of Comeld-2's optimized design, manufacturing trials, assorted mechanical and environmental testing, and a cost assessment comparison to other hybrid joining options available to date.

4.1 INTRODUCTORY STUDY

The primary efforts of the introductory study included:

- Devising of math models and computer simulations of the structural behavior of the Comeld-2 joint undergoing assorted loading exposures
- Alteration of the protrusion pattern based on analytical evaluation of Comeld-2's structural performance
- Development of a manufacturing procedure suitable for outfitting large hybrid structures with the Comeld-2 joint
- Manufacturing trials relevant to both EB metal protrusion and Comeld-2 processing, and fabrication of full-scale Comeld-2 test articles
- Design and execution of a broad test program aimed at experimental evaluation of the joint's structural performance under assorted force-ambient exposures and verification of the validity of selected math models
- Compiling of experimental data and forming of a performance envelope sufficient to specify design allowables for the Comeld-2 joint
- Tweaking the Comeld-2 joint "champion" configuration vested with the best set of service properties
- Preliminary design of hybrid full-scale large-size technology demonstration grillage panel (TDP) with incorporated Comeld-2 joint, replicating an extended material transition structure of a hybrid hull.

The primary aspects of the introductory study presented in this section comprise preliminary finite element (FE) computer modeling and simulations followed by manufacturing trials and mechanical testing. Together they allow for verification of the feasibility and evaluation of the structural efficiency of the newly developed hybrid joining concept. Impact resistance and reparability aspects of Comeld-2 within a hybrid hull were also examined in the introductory study and are presented in the following sections of this chapter.

4.1.1 Initial Computer Simulations

The FE modeling and simulations were carried out for in-plane tensile loading of two types of hybrid joints, a composite double-lap bonded-pinned Comeld-2 and an outwardly identical plain adhesively bonded joint (used as a baseline). The purpose of this exercise was to simulate the pulling of the composite middle plate out from the metal overlaps to determine how the protruding features were incorporated and engaged with the composite affected structural behavior of an adhesively bonded joint.

Two similar heterogeneous parametric FE models comprising second order solid elements, isotropic and orthotropic, of the joint metal and PMC components were built, debugged, and run, employing ANSYS FE software.

The layout of the joint in Figure 4.1 illustrates the employed Comeld-2 configuration with the initially selected protrusion pattern.

A composite middle plate, with thickness $t = 12.7\,\mathrm{mm}$, and two metallic lap plates, with thickness $t_p = 6.4\,\mathrm{mm}$, were chosen to constitute each of the computer models. The envisaged protrusion of the metal plates from the Comeld-2 joint was modeled with idealized tiny upright cylindrical pins, the dimensions and spacing of which replicated a protrusion pattern selected from the array of provided options. Figure 4.2 replicates the image of the initially selected protrusion pattern, presented among the other protrusion options pertaining to the patent (Dance and Kellar, 2010).

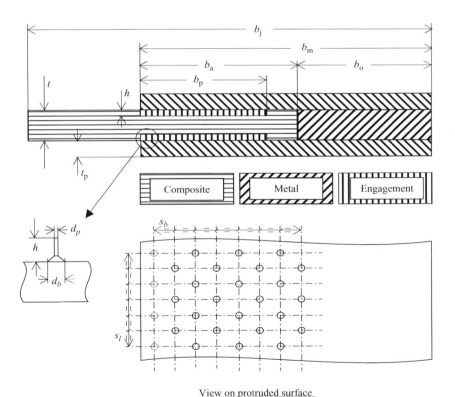

View on protruded surface

Figure 4.1 Comeld-2 layout, adapted from Khodorkovsky and Shkolnikov (2010).

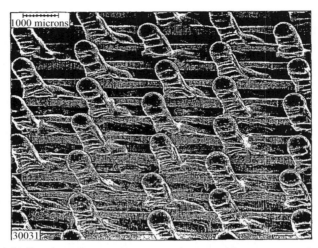

Figure 4.2 Initially selected protrusion pattern, borrowed from Dance and Kellar (2010).

Specifically, a pin's diameter $d_p = d_b = 1.0$ mm and height $h = 3.2$ mm are set. The pins are placed in a staggered order, and the distance between the pins' adjacent rows is set to be $s_l = 3.2$ mm along the joint line and $s_b = 3.8$ mm across that. The thickness of the bonding film is set to be in the range of $0 \leq t_f \leq 0.5$ mm. This implies the presence and relevance of the bonding film to two locations, at the interface of the base flat surfaces of the composite and metal adherends and also at the pin interface with the surrounding composite laminate of the bonded-pinned joint model.

The thickness variation of the bonding film as well as a few other design features were brought in to evaluate their influence on Comeld-2's structural performance. Both adhesive spew fillet at the open ends of the bond-lines and backward bevels of the metal lap plates at their tip were also introduced as optional design features. Additionally, the mechanical properties of the bonding film were varied to evaluate their significance in the joint's performance.

Mechanical properties of the adherend materials used for the computer simulation corresponded to the selected dissimilar material systems certified for marine/naval structural applications. For metal plates, including the protruding pins, the shipbuilding EH–36 steel alloy (ASTM A945 Grade 65), a commercially available analogue of the HLSA 65 alloy, was used for its chemical composition and mechanical properties. The middle plate was composed of E-Glass 24–oz, woven roving with vinyl ester resin (Derakane

8084). The laminate schedule consisted of plies with alternating rotations of 0°, 45°, 90°, and −45°, evenly distributed throughout the thickness of the laminate.

A linear model, free from lateral constraints, was used to reflect the boundary conditions intrinsic to the intended tensile testing. Due to structural and loading symmetries, a half-joint model was utilized.

It was presumed that any structural component of the Comeld-2 joint might embody the weakest link, capable of initiating stress redistribution over the joint body, ultimately leading to its fatal failure. In general, a variety of distinct initial failure modes may occur prior to joint failure. These might be rupture or delamination of the composite middle plate beyond the material transition; de-bonding at the composite-metal interface; fracture of the composite plate in the area of structural discontinuity, e.g., at the tip of the metal overlap or near a pin insertion into the composite plate due to anticipated stress concentration; or metal cracking within the weld and/or other stress concentration areas of the metal part beyond the material transition.

Due to the envisaged diversity of potential failure modes, an integral criterion needed to be applied to reconcile individual failure criteria related to dissimilar joint components with the given design and operational requirements.

An integral criterion needs to be applied to reconcile individual failure criteria of the joint's dissimilar components against the given design and operational requirements.

An integral failure index Ψ_{max} was introduced to satisfy this demand and allow for integration of non-dimensional partial failure indexes $\Psi_M, \Psi_C, \Psi_B,$ which each characterized the stress intensity in a structural component of the joint in relative terms, with respect to the ultimate strength of that component. Accordingly, the integral failure index Ψ_{max} was expressed as

$$\Psi_{max} = \max\{\Psi_M, \Psi_C, \Psi_B\}_V \qquad (4.1)$$

Here, the subscripts "M", "C", and "B" denoted the relevance of partial failure indexes to metal, composite, and bonding film, respectively.

In general, a hybrid joint and its components experience evident three-dimensional stressing under an arbitrary loading exposure. Because of this, the criterion Ψ_{max} reflected both the three-dimensional stress state of the

joint components and the orthotropy of mechanical properties inherent to the utilized PMC. To meet this premise, two conventional failure criteria with the requisite capabilities were employed. These are the von Mises yield criterion, pertinent to the isotropic metal, and the extended Norris-McKinnon criterion, relevant to the nonmetal parts comprising the composite middle plate and bonding film.

The failure index Ψ_M representing the von Mises criterion in its non-dimensional form was expressed as

$$\Psi_M = \frac{\sqrt{(\sigma_1 - \sigma_2)^2 + (\sigma_2 - \sigma_3)^2 + (\sigma_1 - \sigma_3)^2}}{\sqrt{2}\sigma_y} \tag{4.2}$$

where $\sigma_1, \sigma_2, \sigma_3$ were the principal normal stresses being induced within the metal parts under a unit load; and σ_y was the yield stress of the utilized metal alloy.

The extended version of the Norris-McKinnon criterion was chosen because it represented a straightforward option most suitable to characterize the ultimate three-dimensional stress state of a laminate composite with slender orthotropy (or transverse quasi-isotropy) intrinsic to marine-grade PMCs typical for ship structure application. The pertinent failure indexes Ψ_C, Ψ_B were expressed as follows.

$$\left.\begin{array}{c}\Psi_C \\ \Psi_B\end{array}\right\} = \sqrt{\sum_{k=1}^{3}\sum_{l=1}^{k}\tilde{\sigma}_{kl}^2} \tag{4.3}$$

Here $\tilde{\sigma}_{kl} = \frac{\sigma_{kl}}{S_{kl}}$ are relative stress components along the axes of orthotropy of a PMC ($k, l = 1, 2, 3$), also pertinent to a unit load application.

Depending upon a sign of induced normal stress σ_{kk}, resulting from either tension or compression, different ultimate stress values were used:

$$S_{kk} = \begin{cases} (S_{kk})_c & \text{for } \sigma_{kk} \leq 0 \\ (S_{kk})_t & \text{for } \sigma_{kk} > 0 \end{cases}, \quad k = 1,2,3 \tag{4.4}$$

where subscripts "c" and "t" indicate the relevance to the compression and tension stressing, respectively.

Subroutine code based on the given algorithm (4.1)–(4.4) were devised, debugged, and incorporated into the ANSYS FE post-processing procedure. Thanks to this upgrade, the performed computer simulation provided integral characterization of the ultimate stress state of the multi-component

hybrid joint, with visualization of the computed integral failure index Ψ_{max}, allowing for grounded selection of the optimal joint design parameters. Shkolnikov et al. (2009) demonstrate a typical image ensuing from the performed FE simulation displaying distribution of the integral failure index Ψ_{max} over a joint cross section with regard to both hybrid joining options, the conventional plain adhesive bonding baseline and its Comeld-2 analogue reinforced with metal upright pinning.

As anticipated, the area of the metal tip experienced intensified stressing, governing joint performance in both its embodiments. To overcome this unfavorable stress concentration, the modeled metal lap plates were tapered toward the composite middle plate, which noticeably lowered the stress level at the area of the metal tip.

Largely, the metal plates of the bonded–pinned joint underwent considerably more intensive stressing than the bonded–only joint upon achieving the same ultimate stress level at the metal tip area. Apparently, this reflected the influence of the pins' presence, which promoted load–sharing between the metal and composite parts, enabling increased joint load-bearing capability.

Examination of the significance of assorted design features allowed for some level of control of Comeld-2's performance and for preliminary optimization of its design parameters. In particular, it was found that, for given grades and sizing of the metal and composite counterparts, extents $b_p = 102$ mm and $b_a = 152$ mm (per Figure 4.1 notations) in the areas of pinning and entire adhesive bonding across the joint were sufficient for effective metal-to-composite structural engagement. The further increase of these extents, while fairly useful, was not sufficiently effective and might be rejected.

Overall, the performed computer simulation reported in Shkolnikov et al. (2009) did provide valuable qualitative and quantitative data on joint structural behavior helpful for understanding the service interaction of bonded–pinned joint components and grounding of the pursued optimization of Comeld-2 design. Nevertheless, despite a certain sophistication of the employed computer model, the computational results should not be considered conclusive for determination of the ultimate load-bearing capability of a hybrid joint. The reason for this reservation lies in the multiple uncertainties intrinsic to a real multi–component hybrid structure which may noticeably affect the computed result and upset the ultimate performance of that structure.

> Despite a certain sophistication of the employed computer model, the gained computational results should not be considered conclusive for determination of the ultimate load-bearing capability of a hybrid joint because of multiple uncertainties intrinsic to the hybrid joint structure.

Versatility of the integrity, thickness, and mechanical properties of the bonding film, which cannot be dependably determined by implementing only computer modeling, are among the meaningful uncertainties attributive to the large hybrid structures of interest. The preliminary fabrication trials and mechanical testing of full-scale joint articles were carried out to alleviate this concern and gather experimental data to support optimization of the Comeld-2 design.

4.1.2 Manufacturing Trials

To achieve this goal, the joint test articles were designed and fabricated utilizing the above-specified marine-grade materials. Two metal lap plates were used with the initially selected protrusion pattern (see Figure 4.2), each from one surface, then engaged and bonded with the composite laminate.

The main parameters of the protruding upright pins, corresponding to Figure 4.1, were as follows: height $h = 3.0$ mm; diameter $d_p = d_b = 0.5$ mm; spans $s_l = 3$ mm along the joint and $s_b = 5$ mm across the joint. The overall dimensions of the test specimens, per Figure 4.1's notations, were: $b_p = 102$ mm; $b_a = 152$ mm; $b_m = 305$ mm; $b_j = 457$ mm; $t = 12.7$ mm; $t_p = 6.35$ mm. The metal lap plates of the physical test models were tapered toward the composite middle plate as for the FE models discussed above, to lower the stress level at the metal tip and thus increase the load-bearing capability of the joint specimens.

Five principal consecutive material processing steps were executed to produce the hybrid panels with the incorporated Comeld-2 joint. These steps included:
- Preparation of metal plates
- Preparation of fiber layup
- Dry assembly of the two
- VIP setup
- Resin infusion and curing.

Requisite quality control accompanied each processing step.

To clean the steel contact surface of rust and dirty marks to enable proper quality of the adhesive bond between composite and metal adherends, the metal plates underwent grit-blasting and priming prior to the resin infusion. Medium-size grit was used for the grit-blasting. The metal plates then underwent air-blowing and rinsing with methanol over the pinned surface to remove excess grit. Then, the pinned surfaces were lightly coated with a primer to promote a longer-lasting bond with temperature alteration and exposure to harsh environments.

Conventional VIP, the most expedient close-mold material process for manufacturing of large ship structures (Osborne, 2014), was modified to properly accommodate the pins protruding from the metal lap plates within the composite middle plate. The advantages of the modified VIP over other possible PMC processing techniques also included:

- Alleviation of the fit-up problem related to interface of the large metal and composite counterparts being assembled into a hybrid structure
- Elimination of secondary bonding operations, reducing labor intensity and improving structural performance of the joint
- Enhancing repeatability and reliability of the manufacturing process.

Several preliminary manufacturing trials were carried out to tweak the manufacturing procedure, targeting full accommodation of the protruding pins within the composite laminate and corresponding tight contact between the composite and metal adherends, both necessary for maximum realization of the joint structural potential.

Several 584-mm-long hybrid panels containing Comeld-2 joints, with the other dimensional parameters presented above, were produced. Upon VIP processing, the hybrid panels were post-cured at 82 °C for 4 h and then cut using a water-jet cutter to form discrete 63.5-mm-wide test specimens, and subjected to tensile test loading to failure. Basically, the same VIP procedure was executed to produce hybrid panels of both the Comeld-2 joint and its counterpart, the outwardly identical plain adhesively bonded hybrid joint.

As shown in Figure 4.1, the specimens had flat non-tabbed ends for clamping with a wedge-tightening grip. The extents of the mono-material parts of the composite and metal plates beyond their transition were $b_o = b_j - b_m = 152$ mm, sufficient to facilitate both reliable gripping of the specimens in the testing machine and distance of the gripping from the material transition region to allow for leveling of the stress distribution.

4.1.3 Mechanical Testing

To verify feasibility of the Comeld-2 concept and evaluate Comeld-2's load-bearing capability relative to other existing hybrid joining options, preliminary mechanical comparative testing was carried out. As emphasized earlier, the Comeld-2 configuration was specifically selected to enhance its performance under prevailing transverse bending associated with both major categories of operational loading, regular repetitive and dynamic impact/shock loading exposures.

> The Comeld-2 configuration is specifically selected to enhance its performance under prevailing transverse bending associated with both major categories of operational loading, regular repetitive and dynamic impact/shock loading exposures.

Due to this, the bend testing conventional for hull structures was not capable of providing representative data on the ultimate state of the Comeld-2 joint. For this reason, static in-plane tensile tests to failure were executed to signify the most critical load case for the selected metal double-lap hybrid joint configuration. Outwardly identical plain adhesive bonds without metal protrusion were also tested under the same conditions to provide a baseline comparison.

The Tinius Olsen Universal Test Machine with a loading capacity of 534 kN was employed. Largely, the tests were run corresponding to requirements of the ASTM D3039/D3039M standard, except for requirements pertinent to specimen shape and sizing.

The joint test specimens were placed in wedge grips and monotonically loaded in displacement control upon separation of the metal and composite counterparts. The tensile load, strains, and elongation were concurrently recorded and the failure modes and location were noted.

As mentioned in Chapter 3, the baseline plain-bonded joints failed under linear force $F_U = 1944 \pm 35$ kN/m, whereas the bonded-pinned Comeld-2 joints failed under $F_U = 2854 \pm 193$ kN/m. Thus, the initially selected Comeld-2 joint turned out to be 48% stronger than the baseline bonded-only joint with regard to static in-plane tensile loading. The variation coefficients of the test results for the bonded only and bonded-pinned joints were $CV_B = 2.2\%$ and $CV_P = 6.8\%$, respectively, that is consistent with typical variation of results of PMC coupons testing.

Outwardly, all tested specimens exhibited the same failure mode—double-shear of the composite plate being pulled out from the capture of the metal lap plates. Meanwhile, the following fractographic analysis revealed a considerable difference in failure modes of the two tested joint configurations. As demonstrated in Shkolnikov et al. (2009), the bonded-only joint exhibited mainly cohesive failure typical for a well-performed adhesive bonding. The failure mode of the bonded-pinned joint was mixed. Partially, within the pinned area, there was cohesive failure accompanied by either sheared-off or plastically deformed pins. However, in contrast with the plain-bonded specimens, there was predominantly adhesive failure of the bonding film beyond the pinned area of the composite-metal interface. This indicates relatively poor bond performance and hence a decreased contribution to shared load-bearing.

Although the experimental data pertain to a preliminary, non-optimized Comeld-2 technology design option, they convincingly validate feasibility of the novel Comeld-2 joining concept and structural superiority therof compared to baseline plain adhesive bonding. The performed tests also testify to a good match of the selected and employed FE model with the real hybrid joint structure and its ability to reflect specificity of service behavior and failure of the Comeld-2 hybrid structure.

Along with the largely positive outcome, the results of the introductory study revealed certain imperfections inherent to the initially employed Comeld-2 processing. In particular, accommodation of the protruding pins within the composite laminate turned out to be incomplete. This is considered the main cause of the relatively poor performance of both the adhesive bonding beyond the pinned area and the pins themselves. This observation is based on the fact that the plain adhesively bonded specimens, fabricated using the same material components and processes (excluding the protruding pins accommodation), did manifest full involvement of the bond film in load bearing. Because of this experimentally revealed discrepancy, it was believed that a significant increase of Comeld-2's load-bearing capability was attainable by enhancing pin resistance to the lateral force and increasing bonding film involvement in joint performance.

Essentially, the goal was to provide a proper balance between pin robustness and penetrability into and accommodation within the composite laminate. This was the most critical and challenging aspect of Comeld-2 technology, particularly for the tough laminate intrinsic to a marine-grade structural PMC.

Proper balance between the pin robustness and penetrability and complete accommodation within the composite laminate is the most critical and challenging aspect of the Comeld-2 technology, especially as this applies to the tough laminate intrinsic to a marine-grade structural PMC.

Nevertheless, it seemed feasible to properly adjust the initially devised material processing, thereby reaching the pursued synergetic interaction of both joining components, the bonding film and pinning, to substantiate their full engagement and load-sharing. The obvious targets to pursue for realization of this exigency were properly adjusted shape and sizing of the protruding pins along with tweaked material processing that facilitated full accommodation of the pins within the composite laminate. The following development addressed the revealed imperfections of Comeld-2 technology design and provided proper Comeld-2 maturing.

4.2 IMPACT RESISTANCE

Prior to the next steps in Comeld-2 technology design advancement, another discrete aspect of Comeld-2's serviceability was examined in the introductory study. This related to Comeld-2's resistance to impact loading, one of the principal load cases of naval vessels. The relevant efforts, reported by Khodorkovsky et al. (2009), comprised both computer simulations and physical experiments run respectively for three distinct material systems, Comeld-2 composite-metal composition and two control mono-material specimens, fully composite and fully metal.

The employed computer models embodied mono-material and hybrid Comeld-2 plates with fixed boundary conditions at both lateral ends. The width of all models was 305 mm. The thickness of the metal plates was 6.4 mm and that of the composite plates was 12.7 mm for all models. One-quarter symmetry for mono-material models and one-half symmetry for the hybrid joint model were used.

ABAQUS/CAE software was employed to build FE models with chosen boundary conditions. The metal plates were meshed using three-dimensional linear reduced integration elements (C3D8R), whereas the composite plates were meshed with a three-dimensional reduced integration continuum shell element (SC8R) with five-ply layers having the same laminate properties as the net composite.

Approximately medium–high strain rate material characteristics were fed into the FE models to reflect properties changing along with loading rate alteration.

The impactor, confined to moving in the horizontal direction, was assumed to be a 50-mm–diameter hemisphere with varied mass, up to 91 kg, allowing for the desired impact energy. Loading was produced by virtually dropping the impactor from varied heights, up to 1.83 m.

The impactor was also modelled to properly represent its interaction with plates subjected to impact. It was meshed with C3D8R elements in the "no yield" condition.

ABAQUS/Explicit was used to simulate the transient response to the impact loading. In order to predict the onset and track crack development in the metal plates, both damage initiation and evolution values were included in the simulation.

Impact velocities of 4.88, 5.46, and 5.99 m/s were predetermined corresponding to the presumed drop heights of the intended physical experiments: 1.22, 1.52, and 1.83 m.

The 22.7-kg impactor mass at 5.46 m/s impact velocity, producing 338 N-m impact energy, set for the metal model, caused moderate damage to its distal surface. The mass was then increased to 45.4 kg with the same velocity and the damaged area grew significantly, giving crack propagation beyond the acceptable accuracy for predictive behavior.

The applied simplified computer model was deemed sufficient to identify damage onset within a small localized region and did not allow for elimination of supposedly damaged FEs ensuing from load bearing. Due to this, the computed result of a 22.7-kg drop mass was considered to be a threshold value for the onset of damage to the fully metal plate.

The fully composite plate model was subjected to the same impact of a 22.7-kg mass at a 5.46 m/s velocity. Utilizing the Tsai–Hill failure criterion it was found that at least 20% of the distal (strained) surface of the composite plate would exceed the material's load-bearing capability. Based on these findings, it was assumed that both mono–material models, the 6.4-mm-thick metal and the 12.7-mm-thick composite, possessed nearly equal impact resistance for the applied boundary and loading conditions.

In the hybrid joint model, the interface between the metal and composite was simulated by numerically tying together the nodes such that the displacements were linked at the boundary. This approximation assumed that the two surfaces were perfectly bonded and no damage could be assessed at the interface. However, stress conditions in the bonding material could be analyzed to estimate whether bond failure was an issue.

Under an impact mass of 22.7 kg and a slightly increased impact velocity of 6.0 m/s, due to the 1.83-m drop, the hybrid joint model did suffer minor damage. The composite strained surface consumed roughly 50% of its load-carrying capability in the region just outside the material transition. Therefore, the higher impact mass of 55.4 kg at a velocity of 6.0 m/s, producing an impact energy of 816 Nm, was applied. The doubled impact mass notably increased the damage of the composite outside the joint to nearly 90% of the crack threshold. However, it was still localized within the outmost fibers.

The metal overlap plate in turn also suffered tolerable ductile damage of approximately 50% of the degradation threshold under the impact of the 55.4-kg mass. Therefore, both the composite and metal counterparts that constituted the hybrid joint should readily survive the applied 816-Nm impact, which is 2.4 times the impact energy of 338 Nm sufficient to damage either 6.4-mm-thick metal or 12.7-mm composite plates.

Due to the relatively small strained area, the damage would most likely be localized within the directly impacted area and perhaps at the transition edge of the composite, at the metal tip. Overall, the performed impact simulation indicated that Comeld-2 joint should survive a low velocity 816-Nm impact. The impactor would need to have its weight and/or its velocity increased to produce impact energy above the 816-Nm level to induce damage onset at the composite just outside of the metal-composite interface.

Along with the presented computer simulation, five 305-mm-wide test articles replicating the computer models were fabricated and underwent physical impact exposure in the USNA's testing lab (Khodorkovsky et al., 2009). Three of those were hybrid, furnished with Comeld-2 joints, and two others were mono-material control specimens, one made of steel and another of plain composite. The same material systems as those described in Section 4.1 above were used.

The Instron Dynatup 9250HV Impact Test System, suitable for a wide variety of applications requiring low to high impact energies, was employed to execute the testing. The hybrid joint test articles were clamped at the ends of the mono-material extension plates, leaving the other two sides unsupported. A 25.4-mm-diameter hemispheric impactor was used, about that chosen for the computer simulation. A constant 29.6-kg mass was applied, while the impact energy was varied in a range of 271 to 1190 Nm to identify onset failure and its development. A pneumatic brake was used for most tests to avoid repetitive strikes. The contact force, impact energy, and maximum deflection, as well as failure damage extent and location relevant to each test were recorded.

The gathered impact test data presented in Khodorkovsky et al. (2009) perfectly match the computed data outlined above, testifying to the superior performance of Comeld-2 under impact loading when compared to neighboring mono–material structures, either metal or composite.

> The data from impact tests convincingly testify to the superior performance of Comeld-2 under impact loading when compared to that of neighboring mono-material structures, either metal or PMC.

4.3 REPARABILITY

Reparability represents one more major concern relevant to serviceability of a hybrid hull structure that was addressed in the introductory study. Naval vessels side by side with regular operational loading are routinely subjected to accidental and weapon effects that may significantly exceed the loading exposures of regular seagoing operations. Due to this, a possibility of a damage needs be taken into design consideration along with the requisite strength reconciliation for probable damage modes of a hull structure, conceivable aftermath of the acquired damage, and reparability of the hybrid structure in–service.

> The design needs to take into consideration probable damage modes of a hull structure, conceivable aftermath of the acquired damage, and repairability of the hybrid structure while in service.

The diversity of operational loads and potential weapon effects implies a wide variety of possible damage modes for a hull structure. The following generalize the envisaged damage modes referred to in Anon. (n.d.) which might threaten a ship's survival or weaken her combat value.

• Large holes in the underwater hull
• Small holes and cracks in the underwater hull
• Holes in the hull above waterline
• Punctured, impaired, buckled, or distorted bulkheads
• Impaired or ruptured beams, supports, or other structural members
• Ruptured or weakened decks
• Broken or distorted foundations under machinery.

Some damage cases bring inevitable loss, whereas others might be associated just with minor flooding. The vast majority of cases lie between these

extremes. They include the extent of blast damage, fragment holes, warped decks and bulkheads, sprung closures, and ruptured piping, which may or may not be sufficient to produce slow flooding beyond the immediate damage area that sinks the ship.

After damage, the most important factor that determining ship survivability is the ability of her crew to halt progressive flooding by making emergency repairs, e.g., by plugging, patching, and shoring.

The experience gathered over decades of naval operations suggests that after a ship is hit heavily, suffering damage that involves significant flooding, one of two situations usually occurs:

- The damage is so extensive that the ship never stops listing, trimming, and settling in the water, and she goes down within a few minutes after being hit.
- The ship stops heeling, changing trim, and settling in the water shortly after the initial damage.

Moreover, experience shows that in the second case, lasting several hours after damage, the vessel sinking is also possible and the cause of that is directly traceable to progressive flooding (NAVEDTRA, 2008).

Hence, prompt post-damage recovery in-service may become a critical factor for survivability and combatant capability of a warship. And to mitigate the risk of extensive irreparable damage to a hull structure, proper design measures should be undertaken, minimizing the severity of possible damage and providing accessibility for repair.

As a rule, a hybrid hull is structurally more complex than an ordinary mono-material hull, which may bring some extra complications to repair procedures in the event that a hybrid structure in its material-transition region has acquired considerable damage. Nevertheless, the existing reparability requirements pertaining to conventional mono-material hulls should not be diminished with regard to hybrids. To meet this demand, damage proneness of the most responsible hybrid structural components should be minimized, with good reparability and accessibility provided.

At times when the Comeld-2 configuration with outer metal plates embracing the composite has been specifically selected to address this damage-prevention concern it has enhanced damage resistance to a level comparable with that of adjacent mono-material structures. As in the performed computer simulations and physical impact tests, a hybrid panel furnished with Comeld-2 does possess superior, roughly doubled impact resistance compared to that of the control mono-material test articles, either steel or composite. Therefore, as the ship hull underwent impact capable of

producing damage, threatening the ship survival, there was a much higher probability that this would be in a regular monotonic structure, the repair methods for which are well established and relatively simple.

The conventional methods embrace all components of the repair procedure, including nondestructive evaluation of the damage, applying visual and available instrumented methods; common plugging techniques aimed at temporary shutting of holes; and either semi-permanent or permanent repair, allowing for restoration of the damaged structures for a certain period of continued service.

In the low-probability event that the impact is so severe that the Comeld-2 structure also acquires considerable damage, the repair procedure should consist of at least the two following distinct steps. Foremost, temporary patches/plugs available for use onboard most major warships are to be used, exercising conventional patching/plugging routines relevant to monolithic structures to reduce the confronted danger of flooding/sinking. Plate or shaped patches can be welded, respectively to metal region, or bolted/screwed in place, pertaining to either metal, heterogeneous or mono-composite parts of the hybrid structure. Also, other improvised means, such as mattresses and pillows, can be used, along with specially destined patches and plugs.

Another step, succeeding the temporary patch-up, is implementation of permanent repair of the damaged hybrid structure in-service.

Overall, the envisaged damage may comprise de-bonding of the composite from the metal; delamination of the composite laminate; fracture of the composite; or puncturing of the heterogeneous joint assemblage. Such damage may present as either a discrete impairment or a combination of assorted damage modes. Depending on the acquired damage and its severity, an appropriate repair option shall be selected. All repair procedures imply utilization of common material processing steps, similar to those for solely composite structures. These include:

- Evaluation of acquired damage extent
- Removal of damaged material accompanied by stepping of the composite layup subjected to repair
- Cleaning and drying of contact surfaces
- Placement of new fibrous material saturated with polymer resin to compensate for the loss
- Fastening of the joint components being restored with woodscrews or throughout bolts/rivets
- Curing the resin, maintaining a dry air environment at the area of repair.

For a severely damaged hybrid structure, repair may require additional processing steps, including:

- Extending one metal lap plate
- Cleaning and priming the contact surface of that extended lap plate
- Placing fiber preform or either dry or wet prepreg
- Mounting another metal lap plate with a properly cleaned and primed contact surface, followed by the three final steps, as above, for any common repair option.

The outlined repair procedures are considered practical and capable of either fully restoring a damaged Comeld-2 hybrid structure or at least providing long-lasting repair. Meanwhile, it is understood that an in-service repaired hybrid joint may have somewhat reduced load-bearing capability, which should be tolerated. Such indulgence is justifiable because a hybrid joint being repaired does not need to serve for the entire length of a ship's initially assigned life but, rather, is to stay intact for her remaining length of service or until the next docking, where the structure can be fully restored, as necessary.

> While a hybrid joint repaired in-service may have slightly reduced load-bearing capability, this should be tolerated because the repaired structure does not have to serve for the entire length of the ship's life but rather to stay intact for her remaining length of service, or till the next docking, when this structure can be fully restored as necessary.

Repair trials followed by post-repair mechanical testing were carried out to verify feasibility of the devised repair procedures and experimentally evaluate their sufficiency by comparing the load-bearing capability of the original and repaired Comeld-2 test specimens. In particular, two representative joint test articles with acquired distinct failure modes ensuing from the initial destructive static in-plane tension and transverse impact testing were repaired and then underwent the corresponding second, after-repair testing to failure.

Separate counterparts of a joint specimen with failure under tensile loading were sanded and cleaned for secondary bonding. The damaged pins were sanded down. Then, the prepared parts were bonded and bolted with the same resin that was initially used for joint fabrication. Photographs presented in Khodorkovsky et al. (2009) demonstrate images of both the initially

damaged test articles resulting from the first test rounds and those repaired and ready for post-repair testing.

The second, post-repair testing round was also executed at the USNA. As reported by Mouring (2010), the specimen that underwent the tensile loading failed under the linear force $F_U = 1870$ kN/m, and the failure occurred at the specimen's gripping beyond the area of the repaired material transition. This means that the actual ultimate strength of the Comeld-2 specimen being repaired was not less than the experimentally defined value considered the threshold of load-bearing of the post-repair Comeld-2. This was 34% less than the original strength ($F_U = 2855$ kN/m) of Comeld-2, relating to the given configuration above. Apparently, even this understated threshold was satisfactory as provision of full structural restoration was not obligatory for in-service implemented repair.

Due to overlapping of the extra metal plates used to repair the specimen that underwent initial impact testing, a substantial increase of specimen resistance to transverse impact loading was anticipated. This expectation was fulfilled, as the capability of the employed Instron Dynatup 9250HV Impact Test System was insufficient to cause any damage to the exposed repaired specimen, as reported by Mouring (2010).

Overall, the performed examination resulted in convincing validation of Comeld-2's reparability in line with the principal repair procedures outlined above.

4.4 PRELIMINARY ANALYTICAL OPTIMIZATION

Thanks to the proven utility of both principal features of Comeld-2, transverse reinforcement of the composite part with protruding pins and metal double-lap configuration, Comeld-2 possesses superior load-bearing capability under transverse bending compared to ordinary monotonic structures, either metal or composite. It makes the load case, dominant for most hull structures, noncritical for Comeld-2's performance. Due to this, thickness of the composite part did not need to be increased within the material-transition structure customary for hybrid joints. Such design reduction allows for avoidance of local tailoring of the composite part, usually associated with considerable extra labor and added cost.

Essentially, even the decreased thickness (and ply number) of the PMC part within Comeld-2 relative to that of the adjacent ordinary composite panel met the imposed design and operational requirements. Nevertheless,

although allied with potential slight overdesign, keeping the composite plies (thickness) unchanged, leveled with those of the adjacent mono–material panel, makes sense and should be tolerated. The reason for this is the opportunity to reduce the intensity of the related labor operations and to provide an extra safety margin, worthwhile in view of the increased structural complexity of a hybrid joint.

> While decreased thickness (and plies number) of the PMC part within Comeld-2 relative to that of the adjacent ordinary composite panel could suffice the design and operational requirements being imposed, keeping those unchanged makes sense and shall be tolerated to lessen the intensity of the related labor operations and provision of an associated extra safety margin worth in view of the raised structural complexity of a hybrid joint.

While overall advantageous, a number of design parameters of Comeld–2 need to be refined to reduce the imperfections needing to be dealt with and to minimize unfavorable structural peculiarities revealed in the introductory study, thereby maximizing structural performance of Comeld–2. As with any structural component of a ship hull, the material-transition, and the Comeld–2 hybrid joint in particular, need proper structural analysis and strength reconciliation in order to assure robust and verifiable design.

As a matter of fact, the routine analytical procedures of ship hull design are not sufficient to support the requisite structural analysis and design strength reconciliation of a hybrid hull structure. The encountered methodological deficiency is due to both certain distinction of heterogeneous hybrids from conventional mono–material structures and uncertainties inherent to several embodiments thereof. With regard to Comeld–2, the uncertainties primarily pertain to a tangled geometry of the protruding pins and vague mechanical properties of the bonding film at the metal–composite interface.

Due to those considerations, even sophisticated FE software, capable of accurately characterizing structural behavior of a three-dimensional non–linear heterogeneous orthotropic FE model, is not as effective in this case.

The lack of analytical representation of Comeld–2 and its structural behavior noticeably detracts from the efficiency of accurate computer

simulation within a structural design analysis routine. For this reason, simplified math models enabling reasonable accuracy might successfully compete with complex models in the pursued verification of Comeld-2's robustness and its design optimization.

A few simplified math models have been built and run corresponding to conceivable failure modes, including those resulting from the initial mechanical testing to failure. They comprise:

(a)   Rupture of composite laminate weakened by pin insertion at the brink of material transition

(b)   Splitting of the composite-metal interface along with pins shearing off or plastically bending

(c)   Interlaminar shearing of composite laminate beyond the area of the pin insertion

(d)   Adhesive/cohesive failure at the composite-metal interface accompanied by shearing of the composite by the upright pins

(e)   Rupture of the metal lap plates

(f)   Rupture of the weld fillet between the metal middle and lap plates beyond the material transition

(g)   Rupture of the weld seam between the metal middle and extension plates

(h)   Bending of the metal laps outward from the composite middle plate, accompanied by their tearing off.

Sketches in Figure 4.3 delineate these potential failure modes, with dashed lines indicating anticipated crack propagation pertaining to each of the conceived modes.

It is presumed that any of these could occur in rivalry corresponding to the stress intensity induced within a particular joint component, and each of these could initiate a global failure of the joint undergoing an in-plane loading exposure.

Note that, although "g"-mode relates to the failure of a conventional metal-to-metal weld, it is rendered here among other conceived failure options to reflect an opportunity to design Comeld-2 by placing the weakest link beyond the material transition. Apparently, such a design tactic might be preferable for a real structure to avert hybrid joint damage under excessive accidental or combat loading. As reasoned before, although Comeld-2 is well reparable in-service, restoration of its two-/multi-material structure would be a more labor-intensive operation than that for a monomaterial counterpart. Therefore, this probability should be avoided.

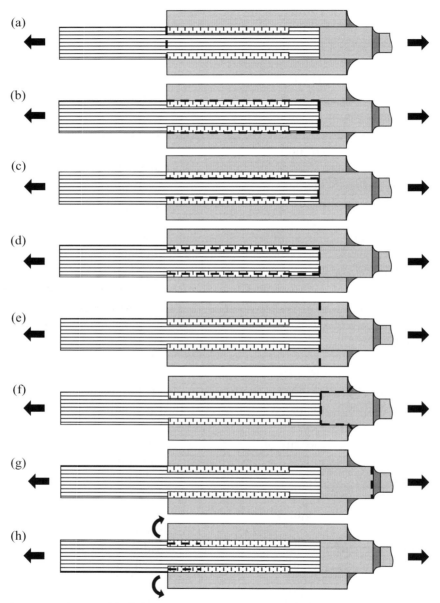

Figure 4.3 Conceivable failure modes, reprinted from Shkolnikov and Khodorkovsky (2011) by permission from SAMPE.

It is presumed that an ultimate force F_{Ui} relevant to the ith load category is sufficient to cause joint failure of the corresponding mode. Therefore,

$$F_{Ui} = \min\{F_{ij}\} \tag{4.5}$$

where F_{ij} is partial ultimate force representing the ith load category that is capable of causing the jth failure mode. The following analytical models are built to define the partial forces F_{ij} with requisite accuracy corresponding to a given set of joint design parameters, properties of utilized materials, and available experimental and/or computed data. The joint design parameters used in the presented models correspond to the notations of Figure 4.1, which outlines Comeld-2's layout.

A couple of assumptions are brought in to bridge input gaps. The first includes a presumption of insignificance of thickness and property variation of the bonding justifiable due to utilization of the VIP-based manufacturing procedure that precludes the fit-up problem and minimizes its influence on joint performance. The other is pin-shape approximation with an idealized cylinder, which also seems acceptable for the undertaken engineering estimates.

The following expressions define the ultimate values of the partial linear forces per joint extent pertaining to the given assortment of conceived failure modes:

(a)   Rupture of composite laminate weakened by pin insertion at the brink of material transition

$$F_a = c_1 \sigma_{tc}\left(t - 2\frac{hd}{s_l}\right) \tag{4.6}$$

where σ_{tc} is the tensile strength of the composite part under pulling force F_a; c_1 is the corrective coefficient that reflects influence of both stress concentration at the composite middle plate near a pin insertion and the pin's actual non-ideal cylindrical shape.

(b)   Splitting of the composite–metal interface along with pins shearing off or plastically bending

$$F_b = 2\frac{\tau_a}{\tilde{\tau}_{am}} b_p \left(\frac{b_a}{b_p} - \frac{\pi d^2}{4 s_b s_l}\left(1 - c_2 \frac{\tau_m}{\tau_a} \frac{G_m}{G_a}\right)\right) \tag{4.7}$$

where τ_a, G_a are the shear strength and modulus of the bonding film, respectively; τ_m, G_m are those pertinent to the metal lap plates; c_2 is the coefficient of stress concentration at the pin base; and $\tilde{\tau}_{am}$ is the peak value of the dimensionless shear stress at the bonding, which is that of a double–lap plain adhesive bond for relevant test data.

The ultimate force relevant to the adhesive/cohesive failure of the plain double-lap adhesive bond is supposed to be

$$F_{bo} = 2\frac{\tau_a}{\tilde{\tau}_{am}}b_a \tag{4.8}$$

(c) Interlaminar shearing of composite laminate beyond the area of the pin insertion

$$F_c = 2\frac{\tau_d}{\tilde{\tau}_{am}}c_3 b_a \tag{4.9}$$

where τ_d is the interlaminar shear strength of the composite; c_3 is the coefficient reflecting the irregularity of the shear stress distribution through the thickness of the composite laminate, approximated as

$$c_3 = 0.8\left(3\left(\frac{h}{t}\right)^2 + 1\right) \tag{4.10}$$

(d) Adhesive/cohesive failure at the composite-metal interface accompanied by shearing of the composite by the upright pins

$$F_d = 2\frac{\tau_c}{\tilde{\tau}_{am}}b_p\left(\frac{b_a}{b_p} - \frac{\pi d^2}{4s_b s_l}\left(1 - \frac{\sigma_{cb}E_{cb}}{\tau_c G_c}\right)\right) \tag{4.11}$$

where σ_{cb} is the bearing strength of the composite.

(e) Rupture of the metal lap plates

$$F_e = 2\sigma_m t_p \tag{4.12}$$

where σ_m is the tensile yield strength of the lap plates.

(f) Rupture of the weld fillet between the metal middle and lap plates beyond the material transition

$$F_f = 2\left(c_4\,\sigma_{me}t_p + \tau_a\frac{G_a}{E_m}b_o\right) \tag{4.13}$$

where c_4 is the knock-down coefficient, reflecting reduction of the material strength due to the notch effect in the unfused area of the weld, presumably lowering its quality in the fillet's root; σ_{me} is the electrode classification number (tensile yield strength); τ_a is the shear strength of the bonding film; G_a is the shear modulus of the bonding film; and E_m is Young's modulus of the metal.

(g) Rupture of the weld seam between the metal middle and extension plates

$$F_g = c_4\,\sigma_{mb}t_e \tag{4.14}$$

(h) Bending of the metal lap plates outward from the composite middle plate, accompanied by their tearing off

The "h" failure mode is due to a certain eccentricity of the force application to each of two metal lap plates that create the bending moments applied to each lap plate

$$M_o = \frac{F_h t_p}{4} \tag{4.15}$$

The conceived bending of the lap plates produces transverse tension in both composite laminate and adhesive bonds within the material transition that may result in delamination and/or de-bonding, accompanied by the pins pulling out from the composite middle plate. It implies that as this failure mode occurs all joint components are involved in effective load-bearing, which is deemed the best attainable instance of structural performance being pursued in the undertaken Comeld-2 optimization. Note that this seems possible, as full incorporation of the protruding pins into the composite and tight metal-to-composite contact are both fulfilled.

Initially, the h-failure mode was taken from the FE simulation briefed in Section 4.1. However, that mode was not observed in the result of the initial Comeld-2 testing round. Apparently, it did not manifest because the pursued full accommodation of the metal protrusion into the composite had not been attained at that stage of development.

Due to the lack of relevant experimental data, the h-failure mode was omitted in the reported preliminary analytical optimization of Comeld-2. On the contrary, the h-mode did manifest and, in fact, dominated the succeeding testing rounds of optimized Comeld-2, which are reported in depth below, in Section 4.5.

All other failure modes and relevant expressions (4.6)–(4.14) have been exploited to characterize performance and provide preliminary analytical optimization of the Comeld-2 design. To do this, expressions (4.6)–(4.14) were fed with relevant data derived from the initial testing round of two sets of joint test specimens, bonded-pinned and outwardly identical plain-bonded joints, as presented above.

Another source of data to feed the given analytical models consists of the mechanical properties of the materials utilized within the joint. Table 4.1 presents such data, comprising properties of the specified steel alloy; results of the composite testing courtesy of the USNA (Mouring, 2009); and estimated values pertaining to all other material properties required for the pursued analytical exercise, based on a few realistic assumptions.

Table 4.1 Estimated mechanical properties, adapted from Khodorkovsky and Shkolnikov (2010)

Materials	Strength, MPa			Moduli of elasticity, GPa		Poisson ratio
	Tension	Bearing	Shear	Normal	Shear	
Composite[a]	323	345	124	20.7	8.0	0.29
Bonding film			107		0.70	0.4
Base metal	448		262	207	83	0.3
Weld metal	483		383	207	83	0.3

[a]Properties across the joint line.

The dimensional parameters of the joint relevant to the initial Comeld-2 design are summarized below, corresponding to notations of the Comeld-2 layout in Figure 4.1:

- Plate dimensions across the joint:
  $b_a = 102$ mm; $b_p = 76$ mm; $b_o = 76$ mm
- Plate thickness:
  $t = 13$ mm; $t_p = 6.4$ mm; $t_e = 11$ mm
- Pin dimensions:
  $d_p = 0.5$ mm; $h = 3.0$ mm; $s_b = 3.7$ mm; $s_l = 5.0$ mm

The computed data obtained from relevant closed-form solutions and numerical FE simulations are also used. In particular, the required parameters of the shear stress distribution along the bond-line are taken from the Volkersen relation, presented in Section 3.2.

Adjustment coefficient c_2 is defined by substitution of the known values of the shear stress $\tilde{\tau}_{am}$ and ultimate loads relevant to both the bonded only (F_{bo}) and bonded-pinned (F_b) joints in Equations (4.7) and (4.8). This, along with other adjustment coefficients c_1 and c_3, is set to match the experimentally determined ultimate forces and failure modes.

So, values of those coefficients are to be: $c_1 = 0.91$; $c_2 = 0.213$; $c_3 = 0.93$, respectively. The c_4 coefficient that is to characterize the stress concentration at the metal weld is assumed to be $c_4 = 0.71$.

Several computing experiments based on the above-outlined algorithm and the estimated values of relevant parameters were run to assess the influence of pin sizing on Comeld-2's performance and provide analytical ground for the improvement of joint load-bearing capability while not compromising its manufacturing feasibility.

The graph in Figure 4.5 demonstrates the computed results, employing expressions (4.6)–(4.14) with regard to the conceived diversity of possible

failure modes; the given parameters of the joint design; and variation of pin diameter in a range of $0 \leq d_p \leq 1.0$ mm.

As can be seen, the pin's diameter meaningfully affects joint performance in more than one failure mode. The other pin and joint parameters as a whole exhibit a similar effect, implying the possibility of a contradictory performance outcome, with either concurrent decrease or increase of partial ultimate forces relevant to distinct failure modes.

For instance, with regard to pin enlargement, joint resistance to shearing off (associated with b–mode) improves, while the laminate's resistance to rupture at the pinned area indicated as the a–mode is lowered, though insignificantly.

Likewise, as pin height is increased, the composite laminate's resistance to interlaminar shearing at the pin tip layers (corresponding to c–mode) grows, whereas resistance of the joint to the a–mode failure is jeopardized.

An increase of thickness of the composite laminate improves joint performance in the a–mode, but its performance pertinent to the b–, c–, and d–modes is decreased.

Thickening of the lap plates improves joint performance for the majority of conceived failure modes, except g–mode, which stays unchanged. However, this improvement is associated with substantially increased weight, proportional to the thickness increase, which lowers weight efficiency of the joint structure accordingly. Extension of the metal overlaps brings similar contradictory results.

The majority of lines representing assorted failure modes in Figure 4.4 are tightly settled. Nevertheless, it seems feasible to control the joint failure mode and location of possible damage by means of design alteration. For a real structure, a damage-prone area should be beyond the heterogeneous material transition, within a neighboring mono-material structure.

Following these tactics, the weakest link of the hybrid joint could and should be positioned beyond the material transition, for instance, at the metal weld, associated with the g–mode, easily reparable in-service.

Contrary to this, one or a few concurrent failure modes pertinent to the heterogeneous part of joint structure should be chosen to represent its ultimate state under test loading.

> In a real structure, the weakest link of a hybrid joint should be positioned beyond the heterogeneous material transition, within a neighboring mono-material structure. One or a few concurrent failure modes pertinent to the heterogeneous part of joint structure should be chosen to represent its ultimate state under test loading.

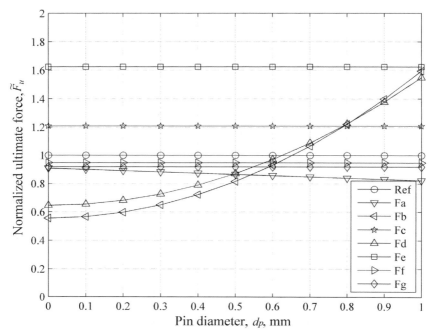

Figure 4.4 Ultimate force versus pin diameter, reprinted from Shkolnikov and Khodorkovsky (2011) by permission from SAMPE.

The graph in Figure 4.5 reflects the computed data for an integral value of the ultimate in-plane tensile force F_U, allowing for overall assessment of Comeld-2's performance as a function of design parameters, such as pin diameter, in particular.

The graph is accompanied by ultimate tension forces pertaining to other basic material and joint design options: monolithic composite plate, extended from the composite middle plate of the joint; the metal weld joint, corresponding to the g-failure mode; and plain adhesive bond. Conjointly, these computed data demonstrate the comparative load-bearing capability of a Comeld-2 joint as a function of pin diameter.

As can be seen, a moderate enlargement of the protruding pins from a diameter $d_p = 0.5$ mm, used for Comeld-2's initial test round, to $d_p = 0.7$ mm, promises a notable ~20% growth of the load-bearing capability from $F_U = 2850$ kN/m to $F_U = 3500$ kN/m.

While this result looks encouraging, it should be kept in mind that pin dimensions, if excessive, impede two vital traits of Comeld-2, full accommodation of the pins within the composite plate and tight contact of that

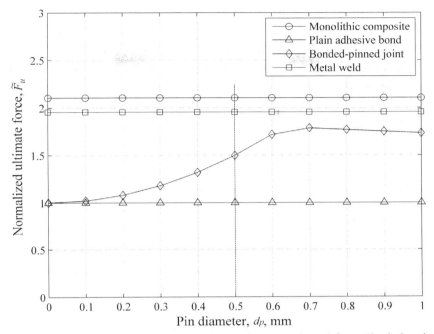

Figure 4.5 Comeld-2 integral performance evaluation, adapted from Khodorkovsky and Shkolnikov (2010).

with the contact base surface of the metal lap plates. A similar contradiction would accompany pin spacing variation. To eliminate these contradictions, pin shape and dimensions need to be cautiously balanced with the permeability of the composite middle plate. Ultimately, as emphasized above, a design that facilitates such a balance would enable maximization of Comeld-2's structural performance.

Altogether, the results of performed computations suggest that the initially selected size of the protruding pins, implemented in the introductory study, is quite close to that of the targeted Comeld-2 optimal configuration. As Figure 4.5 reveals, while a slight enhancement of the pins might be possible, it would not substantially boost joint structural performance. Specifically, a pin diameter of $d_p = 0.65$ mm appears to be the limit to maintain the required balance. Hence, other measures, such pin shape, need to be exploited for further Comeld-2 optimization.

The initially envisioned embodiment of a protruding pin, presented in Figures 4.1 and 4.2 essentially constitutes two main parts, a virtually cylindrical body and its nearly solid cone base. As evidenced by the manufacturing trials,

the conical base noticeably obstructs pin penetration into a tight composite laminate. To alleviate this and ensure full accommodation of the pins within the composite laminate, the shape of the pin base needs be revised.

To meet this demand, tiny radial side stiffeners instead of the solid cone base, such as those demonstrated among exemplified feasible patterns of the patent (Dance and Kellar, 2010), replicated in Figure 4.6, were chosen as a promising option for a pin configuration.

The envisaged pins with radial side stiffeners, while manageably strong and stiff, have much better penetrability than those with a nearly solid cone base, allowing for full engagement of the enlarged pins with the composite, thereby improving the load–bearing capability of the entire joint.

> Pins with radial side stiffeners, while manageably strong and stiff, have much better penetrability than those with a solid cone base, allowing for full engagement of the enlarged pins with the composite, thereby improving the load-bearing capability of the entire joint.

Overall, the performed preliminary analytical study has provided quite valuable qualitative and quantitative input to understanding of the peculiarities of interaction of distinct design features and further improved structural performance of Comeld–2. Based on this, following development of Comeld–2, advanced hybrid joining technology was focused on proper

Figure 4.6 Revised pin configuration, borrowed from Dance and Kellar (2010).

progression of material processing associated with the intended pin reshaping and mechanical–environmental testing of the upgraded Comeld-2, targeting selection of its optimized, "champion" configuration.

4.5 MATERIAL PROCESSING

First and foremost, two–round manufacturing trials were carried out. The first was to tune the EB process to produce a protrusion pattern with the side-stiffened pins. The other was to exploit the chosen protrusion option as a basis for dimensional variation of the pins and verification of accommodation of those within the laminate employed for the composite middle plate, targeting maximization of pin size without compromise, yet with laminate penetration.

4.5.1 Protrusion Trials

Several EB protrusion trials were run to verify feasibility of the devised protrusion pattern for an EH–36 shipbuilding low steel alloy (ASTM A945 Grade 65) of interest. The 1-, 2- and 3-pronged pin patterns, associated with 2-, 4- and 6–leg support, respectively, including those similar to the pattern shown in Figure 4.6, were produced, which quite closely match the looked–for pin configuration, further specified in the patent (Shkolnikov, 2013). Being aware of its feasibility, a few specific targets were imposed to refine the protrusion pattern to maximize of pin resistance to lateral loading, conceived critical for structural performance of Comeld-2 joints:

(1) Further enlarge pin base and height dimensions.
(2) Get a uniformly distributed 6-stiffener (leg) support utilizing the 3–prong pattern, believed to be structurally and penetratively superior to other 1- and 2-prong options.
(3) Keep cavities intruded in the metal lap plates, supplying pin material, relatively shallow and harmless for those lap plates.
(4) Position one of the 3–pin prongs across the lap strip plate to maximize pin bending stiffness against the most efficacious force being applied across the joint.

To be consistent with the targets, the principal dimensions of a prime pin pattern were set which, corresponding to Figure 4.1 notations, were:

- Spacing along the joint $s_l = 5.1$ mm
- Spacing across the joint $s_b = 7.6$ mm
- Pin diameter $d_p = 0.5$ mm
- Pin height $h = 3.0$ mm

- Leg span $d_l = 2.0$ mm
- Leg height $h_l = 2.25$ mm.

The results obtained from the pultrusion trials were for the most part consistent with the specified parameters of pin geometry.

4.5.2 Lap Plate Fabrication

Several $L \times B \times t = 584 \times 305 \times 6.35$ mm blank metal lap plates were protruded with the selected pattern and then underwent their consolidation with the composite layups to form the hybrid panels. The 584-mm length of the metal plates was chosen to meet the capacity of the vacuum pressure chamber employed for the EB metal protrusion.

The extent of the protruding area across the strip plates was $b_p = 102$ mm and that of the entire contact area with the composite was $b_a = 152$ mm, as these values were found in the initial FE simulations to have been sufficient for effective joint structural performance.

Upon implementation of the protrusion, the metal plates were examined for flatness and correspondence of the produced protrusion to the specified parameters. The tolerance grade IT16 (ISO 286) was mainly applied.

Two types of flaws were revealed from the quality examination. One was slight warping of the protruding lap plates caused by thermal distortion due to one-side high-temperature EB treatment. The maximum deflection $w_{\max} \approx 2.5$ mm over the plate's length $L = 584$ mm base was noted. Another imperfection manifested in a few spots of localized deficiency of the protruding pins. Both revealed defects notably exceeded the inflicted dimensional tolerance and needed to be put right to prevent a noticeable negative impact on joint performance.

To do this, the warped plates were flattened during their consolidation with the composite laminate, applying mechanical clamping. Also, spots of deficient protrusion were marked on the even side of the lap plates to identify affected joint test specimens upon consolidation of the metal with composite, with sequential cutting thereof from the formed hybrid panel, to be expelled from the joint specimen clusters.

The protruding plates also underwent other preparatory procedures preceding their assembly and consolidation with the composite. These consisted of grit-blasting with medium-size grit; air-blowing to remove excess grit; rinsing with methanol to clean off any contaminants; and priming with PC-120 primer, supplied by *ITW Plexus*. As directed by *Plexus*, the primer was lightly wiped on with a 25-mm brush and then applied to all

metal surfaces, until obtaining pink cover, indicating sufficiency of the smeared primer.

4.5.3 Composite Processing Trials

The second round of manufacturing trials on processing of hybrid panels was executed prior to fabrication of the test specimens to tweak the resin consistency and distribution within the fiber layup; get the requisite thickness of the composite middle plate being processed; check penetrability of the protruding pins into the fibrous layup; and verify feasibility of the simultaneous formation of the composite laminate and consolidation with the pairing metal lap plates, based on a modified VIP procedure.

As before, in the framework of the introductory study, the composite middle plate comprised a symmetric, balanced layup composed of 20 plies of E-Glass (24-oz, CWR2400, 5×4 woven) fabric utilized as the base part of the fiber layup. The fabric orientation (0°; 45°; 90°; −45°) was altered by plies uniformly distributed throughout the laminate thickness as: ¼; ¼; ¼; ¼ of this.

In addition, a few plies of veil were applied externally and midway at the layup's base part. The external plies were to perform a dual function, to enhance the laminate's resistance to water permeation into the metal-composite interface by applying a resin–rich veil layer, and to ease penetration of the protruding pins into the fibrous material layup. The veil at midpoint of the layup was to match the design thickness of the metal middle plate, that is, 12.7 mm.

Ashland (Derakane 8084) VE-based resin formulation was utilized to make all the hybrid panels: 2% by weight of MEKP hardener and 0.2% by weight of CoNap formulation, with a gel time of slightly above 2 h. This was enough to fully infuse the hybrid panels with the resin. All composite/hybrid panels being processed underwent post-curing in an oven at 71 °C for 6 h.

A modification of the VIP procedure outlined above was devised, providing a proper basis for fabrication of the Comel-2 hybrid panels utilizing metal lap plates with the revised protrusion pattern.

4.5.4 Hybrid panel processing

Two batches of outwardly identical joint test specimens were produced, applying the modified VIP procedure. One represented the Comeld-2 bonded-pinned joint and the other, a plain adhesively bonded joint, used

as a baseline. Six hybrid panels, $L \times B = 584 \times 457$ mm in-plane, were fabricated and split for 63.5-mm-wide test specimens using a water-jet cutter.

Five different configurations of Comeld-2 joint test specimens and one representing the baseline plain adhesive bond, all destined for comparison testing, were produced to various design features related to pin sizing; composite layups; and shapes of the metal lap plates at their tips. The specimens pertinent to the five Comeld-2 configurations were denoted with codes: "1-Aaα"; "2-Baα"; "3-Caα"; "4-Bbα"; and "5-Baβ", whereas the baseline plain adhesive bond was indicated with the code "6-aα".

The first index of the Comeld-2-related codes denoted association with a distinct protrusion pattern, be that "A"; "B"; or "C". The distinction embodied by pin height varied as $h = 2.6$ mm; 3.0 mm; and 3.4 mm, respectively to A; B; and C patterns. Span and height of pin legs were also varied, corresponding to the relations: $d_l = \frac{2}{3}h$; $h_l = \frac{3}{4}h$.

All other pin dimensions were the same for all explored protrusion patterns. Specifically, these included the pin diameter and spacing, which, corresponding to notations of the Comeld-2 layup in Figure 4.1, were: $d_p = 0.5$ mm; $s_l = 5.1$ mm; and $s_b = 7.6$ mm, respectively.

The second index of the code indicated a layup of the composite middle plate. Layup "a" stood for a laminate with external plies of a 10-oz fabric veil (one on each side of the layup) and two plies of mat veil at the midpoint of the laminate. Layup "b" had the same fabric and mat veil plies as for layup "a", all placed externally, one fabric and one mat ply on each side of the layup. The basic part of both layups was the same, as described in Section 4.5.3.

The third index denoted distinctions in the shape of the lap plate edge. The shape option "α" implied a blunt tip, whereas option "β" indicated a backward bevel of 1.3×5.1 mm, which, filled with PMC, was deemed useful in alleviating stress concentration at the metal tip area.

Note that although tapering of lap plates toward the composite middle plate, as employed in the introductory study, did help to lower the peak stress at the metal tip area, as anticipated, the joint specimens for comparison testing presented here were not tapered. The reasons for withholding the tapering were quite pragmatic. First, it was to have a chance to introduce relatively large pins protruding near the tip of the metal lap plates, in order to figure out whether those pins were useful in improving joint performance. Another was to moderate the fabrication cost by eliminating machining of the metal plates, unnecessary for the comparison testing. Another reason was a presumption that tailoring of a real joint, at least of

that destined for a hull's outer shell, would need to be customized to stream-line the hull's external surface, the shape of which should not be replicated in a generic test specimen design.

As regards the lap plate shape simplification was causing notable devia-tion from the initially gained load–bearing capability, strength values derived from the comparison testing and those of the initial feasibility study were incompatible. Because of this, the ultimate strength data from the compar-ison testing round regarding the five Comeld-2 configurations, related to those of configuration 6-aα, represent the newly acquired reference point for plain adhesive bonding.

4.6 CHAMPION SELECTION

The comparison testing of five Comeld-2 configurations was implemented targeting selection of the best, champion option of Comeld-2 design. The testing was followed by further refinement of analytical models capable of reflecting distinct structural behavior of the upgraded joint design.

4.6.1 Comparison Testing

The static in–plane tensile testing to failure of the joint test specimens was executed at room temperature employing an Instron servo–controlled, hydraulically actuated, closed-loop test frame, equipped with a 50-ton load cell and wedge action grips.

The test loading was accompanied by simultaneous determination of time, displacement, and load, at a target rate corresponding to the 100–s length of test loading to specimen failure. The actual rate of the test loading was varied from one specimen to another specimen in a relatively wide range, corresponding to the 71- to 215-s length of loading to specimen fail-ure, in an attempt to get close to the 100-s norm.

To temper the influence of rate variation and provide a level comparison basis, the actual ultimate load data were refined to correspond to the 100-s norm. The "master curve" notion (Miyano et al., 2005; Regel et al., 1972; Shkolnikov, 1995) was utilized for this refinement. This analytical technique implies uniform dependency of mechanical properties and deterioration of PMC on the length of loading exposure, allowing for determination of an ultimate strength value S_s corresponding to a standardized norm of a constant-rate loading by applying the expression

$$S_s = \frac{S}{k_U} \tag{4.16}$$

Here S is the factual ultimate strength determined at the end point of the test loading with a true rate; k_U is a rate adjustment coefficient, which is

$$k_U = 1 - \frac{\ln(\tilde{\vartheta})}{\alpha_0 - 1} \tag{4.17}$$

where $\tilde{\vartheta}$ stands for the ratio of actual length of loading ϑ to the standard loading duration ϑ_s, i.e., $\vartheta_s = 100$ s; α_0 is a dimensionless parameter characterizing time-dependent behavior of a particular PMC composition. The value $\alpha_0 = 32.5$ ensued from introductory fatigue testing of the plain adhesively bonded hybrid joint was used for undertaking adjustment of the ultimate strength values. See Chapter 5 for the grounds and details of this analytical technique.

As anticipated, the load-bearing capability of the Comeld-2 bonded-pinned joint significantly exceeded that of the outwardly identical plain adhesive joint. The best of the Comeld-2 designs was the prime configuration 1-Aaα with moderate pin dimensions and none of the added design features being incorporated into other tested Comeld-2 options. The gained ultimate force pertinent to configuration 1-Aaα was $F_U = 2.57 \pm 0.28$ MN/m versus $F_U = 1.08 \pm 0.19$ MN/m, compared to the baseline plain adhesive bond, configuration 6-aα, demonstrating 2.38× performance superiority in contrast to the 1.48× ensuing from the earlier introductory study.

The other tested Comeld-2 configurations, burdened with add-on optional design features deemed capable of offering some performance gain, although exhibiting substantially stronger performance than that of the baseline, manifested lower structural efficiency than that exhibited by the prime configuration 1-Aaα, ranging from 1.7× to 2.24× compared to the baseline joint performance.

A mixed adhesive–cohesive mode signified the failure of the bonding, as in the initial testing round discussed in Section 4.1. However, ultimate performance of the pins substantially differed from the preceding experience. Contrary to that, when the metal pins were plastically bent or sheared off under the ultimate tensile force, the pins of the newly employed protrusion pattern were dissociated from the composite middle plate, staying upright and sound for all joint loading till its failure.

One more distinctive feature of the newly gained failure mode was the lap plates bending out of the composite middle plate, followed by pulling it away from lap plate capture under the ultimate applied load. Essentially, the

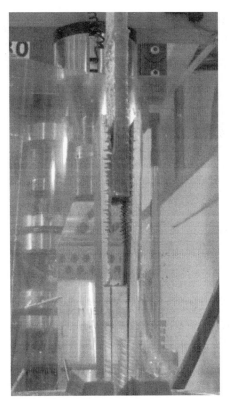

Figure 4.7 Final phase of tensile test, from Khodorkovsky and Shkolnikov (2010).

newly acquired failure mode embodied an earlier predicted result of FE analysis, introduced in Section 4.1. This was then referred to in the list of conceivable failure modes, under index "h," described in Section 4.4.

Figure 4.7 presents the final phase of failure dynamics, illustrating the newly observed failure mode.

Apparently, the revised protrusion pattern was the main cause of the noted transformation of the failure mode. Meanwhile, the implemented pin enlargement seemed to be slightly excessive. While this resulted in pin strengthening, it interfered with tightness of the metal–composite interface which notably affects bond performance.

Hence, taking into consideration all significant aspects of manufacturability and structural efficiency, the simplest configuration 1–Aaα appears to be the right choice for the consequent Comeld-2 champion design finalization.

4.6.2 Upgraded analytical model

The distinct h-failure mode exhibited by Comeld-2 with the newly selected protrusion pattern called for one more analytical model in addition to the assorted models presented already in Section 4.4. Representation of a lap plate as a semi-infinite beam on an elastic foundation loaded with a concentrated bending moment M_0 on the free (tip) end was chosen to characterize joint structural behavior resembling that revealed in the results of the comparison test round.

It is conceived that the bending moment M_0 is formed by the pulling force F_h transferred to the metal lap plates by the composite middle plate via their interface as expressed in relation (4.15). According to Roark and Young (1975), the bending characteristics of the selected model at distance x from the plate edge include:

- Transverse shear

$$V = -M_0\beta\exp(-\beta x)\sin\beta x \tag{4.18}$$

- Bending moment

$$M = M_0\exp(-\beta x)(\sin\beta x + \cos\beta x) \tag{4.19}$$

- Slope

$$\theta = -\frac{M_0}{EI\beta}\exp(-\beta x)\cos\beta x \tag{4.20}$$

- Deflection

$$y = -\frac{M_0}{2EI\beta^2}\exp(-\beta x)(\sin\beta x - \cos\beta x) \tag{4.21}$$

- Foundation stiffness

$$\beta = \left(\frac{lk_0}{4EI}\right)^{\frac{1}{4}} \tag{4.22}$$

- Foundation modulus

$$k_0 = 2\frac{E_c}{c_a t} \tag{4.23}$$

- Bending stiffness of a lap plate

$$EI = E\frac{lt_p^3}{12} \tag{4.24}$$

Here c_a is the coefficient that reflects a difference between the interface area relevant to the pinned joint option and that of the plain–bonded joint; $E_c = 1.0e+3$ MPa is the modulus of elasticity of the composite middle plate in its transverse direction; $E = 2.1e+5$ MPa is the modulus of elasticity of the metal lap plates.

To assess the c_a value with regard to a 3-pronged pin, the external surface of both the protruding pins and the relevant cavities intruding in the metal surface are taken into account. Accordingly, the c_a value is approximated with the expression

$$c_a = 1 + \frac{(2\pi dh + 6d_l h_l)b_p l}{b_a l_s l_{s_b}} \tag{4.25}$$

the parameters of which mainly correspond to the dimensional notations of Figure 4.1; a few others are as follows: l is the joint length (i.e., width of a joint test specimen); d_l, h_l are the span and height of pin legs, respectively.

Figure 4.8 illustrates the experimentally observed phenomenon of a transverse shear force affecting the lap plate with regard to two joint options, the baseline plain adhesive bond and the bonded–pinned joint.

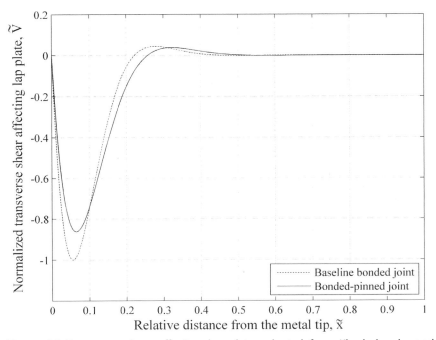

Figure 4.8 Transverse shear affecting lap plate, adapted from Khodorkovsky and Shkolnikov (2010).

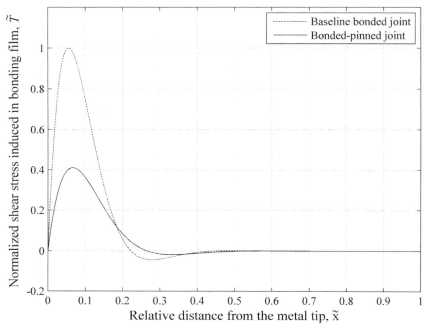

Figure 4.9 Transverse shear stress within bonding film, adapted from Khodorkovsky and Shkolnikov (2010).

Figure 4.9 in turn demonstrates the transverse shear stressing induced by that shear force on the bonding film of those two joints.

As can be seen, the stress state of these two joints is notably different. In particular, the peak stresses induced in the bonding film, which govern the load-bearing capability of the adhesively bonded joints, relate with a roughly 2.4 ratio, which well matches the ratio of experimentally determined ultimate forces of these two joint options. The acquired correlation fairly testifies to the relevance of the devised analytical model to a real Comeld-2 structure in its upgraded form and points to an opportunity to utilize this model for the predictive analysis of Comeld-2's load-bearing capability, at least for comparison, without resorting to utilization of sophisticated FE simulations.

Further, it is presumed that conventional structural analysis routines of a ship hull design can be used to characterize service performance of a material-transition structure being represented as a monotonic structural

component, with a set of reduced mechanical properties being derived from the tests. Moreover, as Comeld-2 is proven be superior to both neighboring mono-material parts, metal and composite, with regard to transverse bending, the conventional structural analysis of the mono-material parts should suffice for design reconciliation of robustness of the entire hybrid structure for this operational load case.

> As Comeld-2 is proven to be superior to both neighboring mono-material parts, metal and composite, with regard to transverse bending, conventional structural analysis of these mono-material parts should suffice for the design reconciliation of the robustness of the entire hybrid structure in this current operational load case.

4.6.3 Champion Joint Design and Fabrication

As emphasized earlier, the distinctive attribute of an effective Comeld-2 design is proper balancing of two contradictory but vital performance traits—complete accommodation of the protruded pins within the composite part and substantial mechanical reinforcement of the adhesive bond by those metal pins. Based on the obtained experimental results, a champion configuration of the bonded-pinned Comeld-2 joint that meets this requirement was selected. Basically, it was a replication of configuration 1-Aaα that exhibited the best structural performance of the comparison joint series, with a superiority of 2.38× that of the plain bond.

Accordingly, a series of 600-mm-long panels with selected design parameters were fabricated and then split for 63.5-mm-wide test specimens of Comeld-2 champion joint. The same materials and largely the same processes as before were employed.

Regarding the manufacture of the comparison joint series, the metal lap plates acquired considerable warping, with deflection up to 5.6 mm along the plate, due to the intense one-sided heating caused by EB treatment, considerably exceeding the imposed requirements for flatness tolerance. To prevent any negative impact of the warping on joint performance, the protruding lap plates were flattened prior to consolidation with the fiber layup. The 5-mm-thick medium-hard rubber slab was used to prevent pins from damage during the flattening operation.

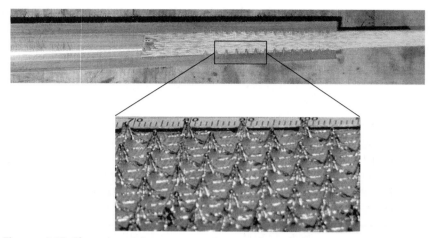

Figure 4.10 Champion joint test specimen, adapted from Khodorkovsky and Shkolnikov (2011).

Two alternative assembly and composite processing procedures were executed. One, resulting in the "1S" specimen series, was accompanied by debulking of one lap plate to the fiberglass layup. Another "2S" specimen series involved debulking of both metal lap plates. The first option, while preferable for implementation with large-structure Comeld-2 technology, was found to lack complete pin penetration into the fibrous material. The second option, in contrast, while it did provide proper pin accommodation within the composite, added complication to the material processing with regard to fabrication of a large hybrid panel.

To figure out whether the potential processing deficiency intrinsic to the 1S-series considerably affected structural performance of the produced joint, both processing options were carried out and both fabricated joint specimens underwent experimental examination.

A total of five 600-mm-long champion hybrid panels were fabricated and then split employing a water-jet cutter into 63.5-mm-wide specimens, which underwent assorted mechanical-environmental and watertightness testing. The photograph in Figure 4.10 demonstrates a joint test specimen cut from a Comeld-2 hybrid panel being produced.

4.7 MOISTENING/WATERTIGHTNESS EXAMINATION

Moistening represents one more concern about the serviceability of a hybrid structure, which has been experimentally evaluated with regard to water penetration into the composite-metal interface of Comeld-2. In general,

moistening may affect the service performance of a hybrid structure in a diversity of ways, via

- Deterioration of PMC within a hybrid structure
- Added stressing due to PMC moisture-originated swelling
- Crevice corrosion of the metal part at the composite-metal interface, accompanied by de-bonding and ultimately premature failure of the entire hybrid structure.

Due to these potential drawbacks, water penetration into the composite-metal interface of the material-transition is prohibitive for a real hybrid hull structure. The chosen Comeld-2 configuration together with the employed co–curing manufacturing procedure was specifically selected for and was capable of providing the requisite watertightness of the interface, ensuring the corrosion-free state of the paired metal. This optimistic assertion was based on longstanding operational experience with combined bonded-bolted joints conceptually similar to the bonded-pinned Comeld-2 structure.

As mentioned above in Section 2.2, bonded-bolted joints were utilized for outboard hybrid structures of several Russian submarines that were in commission for two decades (1970s–1990s). Neither PMC deterioration nor metal corrosion was observed during periodic inspections and no complaints about the service performance of the composite outer hull panels outfitted with the bonded-bolted joints were reported by the crews during the subs' service.

Likewise, Boyd et al. (2004) report good resistance of co–cured joints to long-term aging moisture exposure ensuing from an investigation of the integrity of steel-composite joints used for deck–to–superstructure connections.

Despite this positive outlook, several preventive measures were undertaken to ensure requisite watertightness and corrosion resistance of the champion Comeld-2 specimens destined for moistening and watertightness testing. These included:

- Proper cleaning and priming of the metal lap plates before their assembly and consolidation with the composite part
- Utilization of a marine-grade polymer resin to provide requisite water resistance of the PMC laminate
- Use of a resin-rich veil for the external protective layers of the composite part
- Sealant application at the tip of the metal laps.

In particular, radial fillets composed of polymer resin (Ashland FV 8084 Derakane VE) and 20% by weight of microspheres were provided at the

metal tip on both sides of the hybrid panels upon their consolidation. A thermo-resistant long-lasting, permanently flexible marine polysulfide sealant, usable both above and below the waterline and capable of withstanding 35 °C, was also applied to the cut side surfaces of the joint specimens to protect them from moisture exposure unrelated to the actual operational conditions.

To verify sufficiency of these measures and evaluate resistance of the champion bonded-pinned joint to seawater exposure, the hybrid panels underwent moistening-watertightness examination carried out in two consecutive test rounds. The first was to experimentally define the length of moisture exposure sufficient for the PMC to gain effective water absorption equilibrium. The second was to experimentally examine whether a hybrid joint was capable of withstanding such moistening without formation of rust stains at the metal-composite interface.

The external layers of the composite middle plate were considered to be the only channel for potential water permeation into the metal-composite interface of Comeld-2. For this reason, two different thin composite plates destined for the first round testing were fabricated. One of those was a 4-ply, 1.1-mm-thick plate that represented the external protective layer, composed of two outer plies of E-Glass, a 10-oz woven fabric veil and two inner plies of E-glass mat veil. The other was a two-ply, 1.4-mm-thick plate made of 24-oz E-Glass woven fabric (CWR 2400, 5×4) that embodied the base part of the used fiber layup, presumed to be self-sufficient without any additional protective layer. The same Derakane 8084 VE resin was applied for both material compositions. VIP was employed to process both plates, which were post-cured at 71 °C for 6 h and then each split into five square 100×100 mm material specimens for test moistening.

An immersion tank filled with synthesized seawater was set up, corresponding to requirements of the ASTM D5229/D5229M-92 standard. The tank was furnished with a propeller blade stirrer, a heater capable of maintaining a steady 35 ± 3 °C temperature, and a thermostat to control the temperature throughout the test. The elevated temperature was to imitate the worst possible operational conditions with regard to both seawater absorption and metal rusting.

The precision weighing of the specimens being progressively moistened was carried out employing an Acculab VIC-123 milligram scale. The percent moisture mass gain ΔW_m was monitored and plotted versus time as

$\triangle W_m = \sqrt{t}$ for all test specimens. The mass change $\triangle W_m$ was calculated for a time interval as

$$\varDelta W_m = \left| \frac{W_i - W_b}{W_b} \right| \times 100\% \qquad (4.26)$$

where: W_i is the current mass of ith moistened specimen; W_b is the baseline specimen mass.

Following this routine, it was defined that the effective moisture equilibrium for both tested PMC compositions sets over 64 days was sufficient for the watertightness testing of a hybrid joint panel as well.

The second test round was also executed in correspondence with Procedure B of the ASTM D5229/D5229M-92 standard. Six hybrid joint specimens were maintained in the same steady-state seawater environment. The testing setup, similar to that employed for the first test round, comprised an immersion tank furnished with a heater, thermo-controller, and propeller blade stirrer.

Upon completion of the planned two-month moistening exposure, the joint test specimens were cleaned of sealant on the cut sides and visually examined for any sign of rust stain at the metal-composite interface. This did not reveal any sign of metal rust, thereby validating the anticipated sufficiency of the selected Comeld-2 design technology for watertightness at the composite-metal interface.

The photograph in Figure 4.11 showing the joint specimens that underwent the 2-month seawater moistening convincingly testifies to this assertion.

Owing to the positive experimental outcome, there is no need to incorporate a stainless steel alloy such as AL-6XN into a Comeld-2-based material-transition structure, as might be necessary for other hybrid joining options, to protect the metal part of the hybrid structure from corrosion. This considerably simplifies the targeted implementation of a hybrid hull and decreases the cost of hybrid hull construction and its in-service maintenance.

The Comeld-2 hybrid joint is self-sufficient in preventing water penetration into the metal-composite interface and protecting the metal part from corrosion; hence, there is no need to incorporate stainless steel alloy into the material-transition structure as might be necessary for other hybrid joint options and considerably simplifies implementation of hybrid hull construction and reduces its cost.

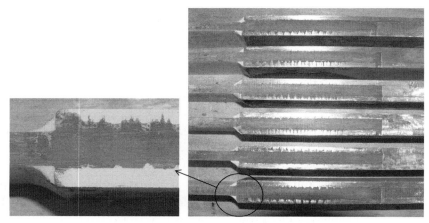

Figure 4.11 Post-moistening examination of joint specimens, from Khodorkovsky and Shkolnikov (2011).

4.8 CHAMPION MECHANICAL-ENVIRONMENTAL TESTING

In-plane tensile testing of the Comeld-2 champion to failure under varied ambient conditions was carried out. The principal test target was to obtain experimental data sufficient to evaluate serviceability of Comeld-2 in a ship hull for the anticipated range of operational exposures.

4.8.1 Test Set-up

The procedure executed for the champion joint testing mainly replicated the ASTM D3039/D3039M standard that is primarily dedicated to determination of the tensile properties of PMCs. The shape and dimensions of the joint test specimens were the same as those utilized for the introductory and comparison test rounds and specified in Section 4.1. The same configuration, with flat non-tabbed specimen ends, clamped with wedge-tightening grips, was also used.

To embrace a representative range of ambient conditions typical for ship hull operation, the thermal influence was evaluated at three temperature levels: lowered, room, and elevated, at $T = -15\,^{\circ}\mathrm{C}$, $23\,^{\circ}\mathrm{C}$, and $60\,^{\circ}\mathrm{C}$, respectively.

The test specimens subjected to testing at non-room temperatures were conditioned under assigned lowered or elevated temperatures prior to the tests for 3 h each. Destined parameters of the ambient conditions were maintained, with deviation from the specified nominal levels not exceeding $\pm 3\,^{\circ}\mathrm{C}$ of temperature and $50 \pm 10\%$ of relative humidity for the dry tests.

For this purpose, a thermally insulated temperature–humidity-controlled chamber was employed. The specimens, having undergone the preceding moistening/watertightness testing, were kept wet till execution of the mechanical testing.

The static test loading was set in force control mode with constant rate $R = 1.8$ kN/s, approximately corresponding to the targeted norm $\vartheta = 100$ s of the loading length to specimen failure. The fatigue tests were run employing a low–cyclic pulse loading with the standard broken–line (triangle) profile. The cycling frequency was set at $f = 0.133$ Hz, a level corresponding to characteristic alternation of ship seaway loading. The load ratio was set to $r \equiv \frac{F_{min}}{F_{max}} = 0.1$.

To gain representative fatigue data, the maximum force F_{max} was set within a range of $0.45 F_U \leq F_{max} \leq 0.8 F_U$, where F_U was the mean ultimate force defined by the static tests executed prior to the fatigue testing.

Onset of separation of the composite laminate from capture of the metal laps was regarded as a failure criterion for the static tests. A crack size of $\sim 10\%$ of the extent of the metal–composite interface was chosen as the failure criterion for fatigue testing.

An MTS Model 880 servo–hydraulic test frame equipped with wedge action grips was employed for load application and specimen restraint. An MTS Model 661.23A-01 load cell with a maximum capacity of 250 kN was applied to generate the test tensile force.

Software designed by WMT&R was used to provide means for simultaneous readings of time, displacement, and load. As before, the master curve technique expressed in formulas (4.16) and (4.17) was employed to unify the test data, corresponding to the selected 100-s norm of test loading to provide a proper basis for test results comparison.

Control measurements of all test specimens were carried out prior to the testing. This did not reveal any specimen width variation exceeding $\pm 0.5\%$ of the nominal width value, and due to insufficiency these dimensional deviations were neglected in the following analysis of the test data.

4.8.2 Static Testing

The test data on load-bearing capability of the champion joint configuration are presented in Table 4.2, including the one- and two-side debulking options, denoted as "1S" and "2S," respectively. Both absolute values of the ultimate test load for standardized length of loading and the relative values of the strength of the plain adhesive bond are shown.

Table 4.2 Summary of champion static test data, adapted from Khodorkovsky and Shkolnikov (2011)

Specimen category	Ultimate linear load, MN/m	Ultimate load comparing to reference				
		Config. 1-Aaα	Plain-bonded joint	Bonded-bolted joint	NJC's adhesive bond	Composite laminate
Comeld-2 champion						
1S	2.34±0.26	0.91	2.2	0.90	1.6	0.57
2S	2.71±0.16	1.05	2.5	1.05	1.8	0.66
References						
Configuration 1-Aaα	2.57±0.28	**1.0**	2.4	0.99	1.7	0.63
Plain-bonded joint	1.08±0.19	0.42	**1.0**	0.42	0.72	0.26
Bonded-bolted joint	2.59	1.01	2.4	**1.0**	1.7	0.63
NJC's adhesive bond	1.50	0.58	1.4	0.58	**1.0**	0.36
Composite laminate	4.10±0.09	1.59	3.8	1.60	2.7	**1.0**

Also, load–bearing capability of other joint configurations and material coupons under in–plane tension are presented here for reference. These include the best Comeld-2 configuration, 1-Aaα from the preceding test round; the baseline plain adhesively bonded joint outwardly identical to the Comeld-2 champion; a bonded–bolted joint, the most structurally efficient hybrid joining option available to date, test data on which are courtesy of NSWCCD (Loup, 2010); the NJC adhesive bond introduced in Section 3.3, test data on which are borrowed from (Brown, 2004); and a material coupon of the composite laminate utilized within the joint, test data of which are courtesy of USNA (Mouring, 2009).

Note that although the bonded–bolted joint presented here has a similar layout (except for the use of bolts instead of pinning) and utilizes the same material grades as those used for the Comeld-2 bonded–pinned joint, contrary to the Comeld-2 this has a double-thick composite middle plate. As the load-bearing capability of a bonded–bolted joint under in–plane tension is roughly linearly proportional to the middle plate thickness, the actual strength value $F_U = 5.18$ MN/m, reported by Loup (2010), is reduced to $F_U = 2.59$ MN/m to render provide a proper comparison basis and make all the compiled test data comparable.

Note also that the load-bearing capability of the composite laminate presented here for reference is not fully applicable to joint structural performance within a ship hull. The applied in–plane tension that represents the worst load case for a butt joint's performance does not reflect a complex loading intrinsic to hull structure operation with the prevailing transverse bending. For this load case, for which the metal double-lap design was specifically selected, the Comeld-2 hybrid panel greatly surpasses performance of an ordinary mono–material panel, either composite or metal. Results of the transverse impact testing of the Comeld-2 joint presented in Section 4.2 testify to this advantage.

As can be seen, the champion Comeld-2 specimens of the 2S-series exhibit superior performance to all joint references, including the strongest existing bonded–bolted joint option. Specifically, $2.5\times$ superiority of the 2S-series in load-bearing capability under static loading is demonstrated comparative to the conventional adhesive bond, and there is an approximately 5% advantage over the strongest existing, but substantially heavier, more labor intense, and costlier state-of-the-art bonded–bolted joint.

The Comeld-2 champion design exhibits superior structural performance to all hybrid joint options available to date, including the strongest existing bonded-bolted joint. Specifically, 2.5× superiority in load-bearing capability under static loading is demonstrated when compared to conventional adhesive bonding, and there is also an approximately 5% advantage over the strongest existing, but substantially heavier, more labor intense, and costlier state-of-the-art bonded-bolted joint.

The 1S–series is notably (16%) lower in performance than the 2S–series. Nevertheless, it deserves to be kept as an acceptable option due to its fairly simpler material processing, which may turn out to be the only option for manufacture of intricate hybrid parts for a real hull structure.

The graph in Figure 4.12 summarizes the results of the ambient–altered static testing of the champion joint for five distinct testing environments: cool temperature dry (CTD); room temperature dry (RTD); room temperature

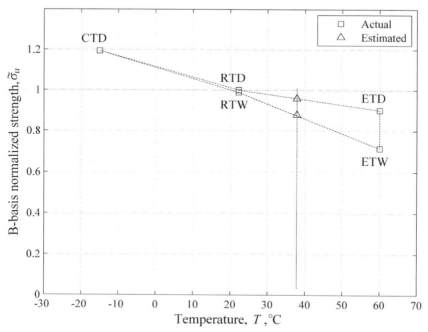

Figure 4.12 Champion performance envelope, adapted from Khodorkovsky and Shkolnikov (2011).

wet (RTW); elevated temperature dry (ETD); and elevated temperature wet (ETW).

Overall, the ultimate strength range of the joint is $0.68 \leq \tilde{S} \leq 1.20$, where $\tilde{S}$ stands for a normalized value of the ultimate strength for the given alteration of ambient conditions, i.e., for the ratio of an experimentally acquired absolute strength value determined at given particular ambient conditions and the strength value under normal testing conditions.

On the whole, the obtained ambient–altered test data represent a performance design envelope allowing for determination of joint load-bearing capability for an arbitrary operational situation. It should be noted that the strength variation corresponds well to that of a plain structural composite due to its governing role in the hybrid joint's load bearing.

The performed champion joint tests have resulted in three slightly different failure modes of individual joint specimens. One of the modes is adhesive-cohesive de-bonding of the metal-composite interface accompanied by the lap plates folding outward from the composite and pulling it from lap capture. Another is delamination of the composite plate itself within the lap. And the third mode observed is a combination of those two ordinary failure modes.

It should be stressed that the discerned variety of acquired failure modes is associated with a relatively narrow range of ultimate strength values. This indicates that the selected pattern of metal protrusion and, hence, metal-to-composite interface of the champion Comeld-2 have practically attained their maximum possible level of performance. Due to this, any further improvement of the joint as a whole should be sought beyond the interface pattern, but rather relatively to the metal tip tapering and/or some extension of the bonding area at the metal-to-composite interface.

The significance of pinning and its interaction with bonding for Comeld-2 load-bearing capability along with the substantial distinctiveness of ultimate strength values is illustrated by the results of the strain-gauging that accompanied the static testing. Two outwardly identical joint specimens, one pertaining to the Comeld-2 champion cluster and another relevant to the plain bond, underwent the strain-gauging, for which *Vishay Micromeasurements* uniaxial general purpose strain gauges, item #CEA-06-250UW-350, were utilized.

The quantitative data obtained on distribution of the strain along the composite-metal interface is presented in Figures 4.13 and 4.14 for the plain bond and champion Comeld-2, respectively.

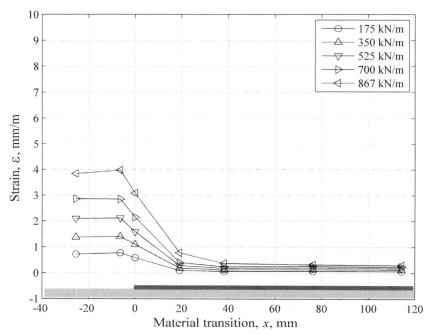

Figure 4.13 Strain distribution along material transition of plain bond, adapted from Khodorkovsky and Shkolnikov (2011).

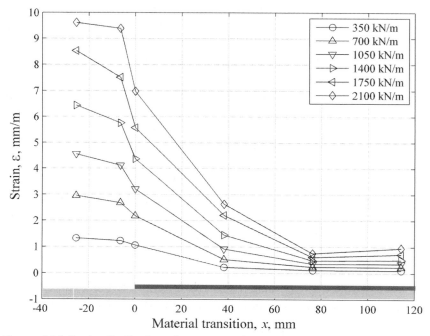

Figure 4.14 Strain distribution along material transition of champion Comeld-2, adapted from Khodorkovsky and Shkolnikov (2011).

The present charts illuminate well the influence of pinning on Comeld–2's performance. The bonded–pinned joint demonstrates intensified involvement of joint material in load–bearing, which substantiates the observed improvement in structural performance compared to the plain bond. It should be emphasized that this experimental result is fully consistent with the outcome of the computer simulations presented in Shkolnikov et al. (2009) and discussed in detail in Section 4.1 above.

4.8.3 Fatigue Testing

Along with static testing, both the champion bonded–pinned joint and plain adhesively bonded joint underwent low-cyclic triangular waveform test loading. The graph in Figure 4.15 plots the acquired fatigue data pertaining to the two hybrid joining options.

Here markers represent dimensionless ratios of the experimentally determined fatigue strength and the ultimate static strength of the champion joint against the cycles count. The actual cyclic test results are added to the graph with ultimate strength data determined under the normal conditions of static

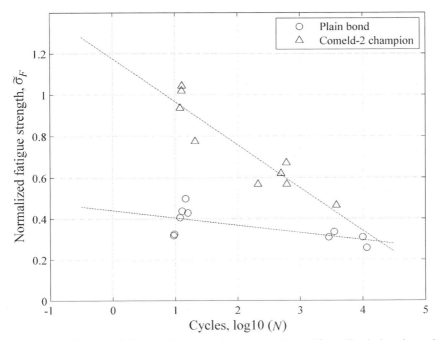

Figure 4.15 Combined chart on fatigue performance, adapted from Khodorkovsky and Shkolnikov (2011).

testing and used here as half-cycle fatigue data. To make the static–origin data consistent with the results of the cyclic loading, those values have been reduced to frequency $f=0.133$ Hz, employing the load rate unification technique based on the master curve notion.

The graph in Figure 4.15 also presents estimated fatigue performance derived from a semi-logarithm approximation of the strength–cycles diagram conventional for fatigue characterization of structural PMCs. Specifically, the following expression regarding normalized fatigue strength $\tilde{\sigma}_F$ is utilized

$$\tilde{\sigma}_F \equiv \frac{\sigma_F}{S} = B_N - \beta_N \log_{10}(N) \tag{4.27}$$

where σ_F is fatigue strength under a given count of cyclic loading; S is the mean ultimate strength at normal static loading conditions; B_N and β_N are fatigue parameters intrinsic to a particular hybrid joint pattern for standard cycling frequency and normal ambient conditions. Specifically, the fatigue parameters revealed in the presented joint testing are as follows:

- For plain adhesive bond: $B_N=1.094$; $\beta_N=-0.089$
- For champion Comeld-2: $B_N=1.176$; $\beta_N=-0.208$

The two experimentally defined strength characteristics, ultimate static strength and fatigue strength versus cycles count, represent the prime determinants of structural performance of a hull structure. Generally speaking, the static strength governs structural efficiency for short-term, episodic, and/or dynamic loading, distinctive to warship operation, whereas fatigue performance is a principal characteristic of hull long-term load-bearing capability pertaining to regular, e.g., seaway, operation.

As can be seen in the graph in Figure 4.15, although the champion joint manifests significantly higher (2.5×) load-bearing capability under short-term loading than the plain bond does, fatigue performance of the two for long-term loading is fairly similar.

The observed disproportion is justifiable in view of the added stress concentration caused by multiple insertions of protruded metal pins into composite laminate. This makes a bonded–pinned joint somewhat more sensitive to fatigue loading compared to the plain bond. It should be noted that the state-of-the-art bonded–bolted joints, conceptually similar to the bonded–pinned Comeld-2, manifest the same phenomenon of lowered fatigue performance. The ordinary bolting in turn exhibits an even higher degree of downgrade.

The fatigue data given in Figure 4.15 essentially represent the performance of the two utmost configurations of a bonded–pinned joint, one with

an optimized protrusion pattern that supplies the feasible maximum of joint load-bearing capability under static loading, and the other with no pinning presence at all. Apparently, there is an opportunity to tweak the metal protrusion and thereby rationally balance joint structural performance, as the highest attainable ultimate strength is not required for a particular joint application. In this case, the pin sizing can and should be somewhat decreased to enable a proportion of joint static and fatigue strength that best fits the joint's particular operational assignment.

> Comeld-2's load-bearing capabilities under static and fatigue loading can be rationally balanced by tweaking the metal protrusion respectively to a particular joint application assignment not requiring its highest attainable ultimate strength.

Figure 4.16 illustrates this notion for the experimental data as a function of the relative density of the pinning δ ranging as $\delta_{min} \leq \delta \leq \delta_{max}$.

Here, density $\delta_{min} = 0$ implies no pinning, for the plain adhesive bond, whereas $\delta_{max} = 1.0$ refers to the maximum feasible number of pins

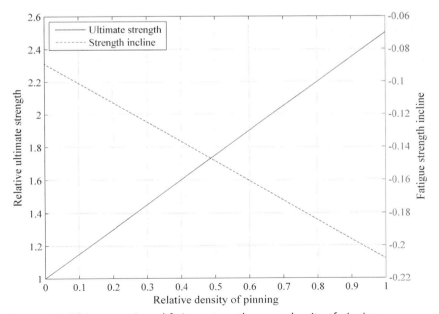

Figure 4.16 Ultimate static and fatigue strength versus density of pinning.

incorporated into the composite laminate of a bonded–pinned joint corresponding to that of the champion Comeld-2 configuration.

For instance, as the doubled ultimate strength of the plain adhesive bond (instead of the maximum available 2.5×) is considered sufficient for a particular Comeld-2 application, it is appropriate to employ pinning corresponding to its intermediate density $\delta \approx 0.67 \delta_{max}$, which is associated with significantly improved fatigue performance, related to a fatigue strength incline of roughly $\beta_N \approx -0.167$, instead of that intrinsic to the Comeld-2 champion ($\beta_N = -0.208$).

Reliability is another critical attribute of structural performance that needs to be evaluated for the serviceability and structural/weight efficiency of a hybrid structure. In this matter, the champion Comeld-2 is notably superior to the plain bond. The coefficients of variation regarding the obtained test data of the two joining options are 5.57% and 17.3%, respectively.

With this outcome, the bonded–pinned Comeld-2 technology comes to the fore as the best hybrid joining option for heavy–duty naval/marine structure application for the entire set of performance parameters for manufacturability and serviceability, including joint load-bearing capability, reliability, weight effectiveness, and maintainability.

4.9 TECHNO-ECONOMIC APPRAISAL

Along with serviceability requirements, cost affordability is one of the principal criteria when considering the practicality and suitability of a new shipbuilding technology. While structurally superior, Comeld-2's manufacturing cost is to a certain extent higher than that of a plain adhesive bond, due to the added metal protrusion operation. This is particularly noticeable at the current research, not-yet-industrial state of Comeld-2 development. However, the robustness and reliability of a plain adhesive bond are substantially less than those of Comeld-2, which typically inhibits its application for heavy–duty naval structures.

Meanwhile, the weight and cost of Comeld-2 are considerably lower than those for both the bolted and bonded-bolted joints common for naval uses. These weight and cost advantages are due to elimination of the weighty bolt-nut components and exclusion of the related labor-intensive hole drilling and bolt-nut coupling operations.

The weight savings of a ship hull structure are essentially convertible into cost savings by Comeld-2's contribution to reduction of the ship's

displacement, enhancement of her payload capacity, lowered energy consumption, and relevant fuel economy, among other advantages. Hence, to provide fair judgment on Comeld-2 affordability, its cost should be assessed by taking into account both the anticipated manufacturing expenses and associated indirect cost savings. Otherwise, an incomplete cost evaluation may ill-advisedly put off use of the progressive Comeld-2 hybrid joining technology for beneficial industrial implementation.

To clarify the issue and alleviate the encountered concern about cost effectiveness and affordability, weight and cost of a conceivable replacement of the ordinary bolted joint with Comeld-2 was estimated. To be specific, the bolted joint currently employed to mount the composite topside structures on the deck of the DDG-1000 *Zumwalt* class destroyer was used as a baseline for the weight and cost assessments. The following algorithm is employed for the rough order-of-magnitude (ROM) estimates.

The anticipated weight savings, ΔW, resulting from the conceived joint replacement for a ship, is expressed as

$$\Delta W = (w_B - w_{C2})L_j \tag{4.28}$$

where W_B and W_{C2} denote weight rates per linear unit of the joint extent for the bolted and Comeld-2 joints, respectively; L_j is the total length of the composite-to-metal seam.

Per Brown's (2004) assessment, the weight rate of the bolted joint utilized for the composite superstructures of the DDG-1000 class destroyer is $W_B = 162$ kg/m, whereas that of the new NJC adhesively bonded joint (Simler and Brown, 2003) is $W_{NJC} = 94$ kg/m. The weight estimate of Comeld-2 that exceeds structural efficiency of either bolted or NJC adhesive joints (Table 4.2) is substantially less. It is $W_{C2} = 29$ kg/m for the champion Comeld-2 configuration.

Due to this weight advantage, the potential replacement with Comeld-2 would result in $\Delta w = 133$ kg/m of direct weight savings per linear meter of joint extent. In the case of use of a Toray T700-based CFRP such as that employed for the DDG-1000's composite structure (LeGault, 2010) instead of the GFRP employed for the Comeld-2, the weight savings would be even more significant.

Depending upon a detailed structural design, the total length of joint L_j needed to attach a composite superstructure to the metal hull might range broadly. According to Anon. (2011) the only found publicly released the seam extent of, the DDG-1000 composite deckhouse, skirted with the metal, requires more than $L_w = 6100$ m of welding to the base metal structure.

Presumably, this should correspond to the length of the metal-to-composite joint. The given weld extent probably implies its aggregated value pertaining to two-sided welding of two metal lap plates used for both composite topside structures, the deckhouse and hangar. As such, the total length of a hybrid joint line is about $L_j = L_w/_4 \approx 1525$ m. While traceable, the estimated total length of a hybrid joint line seems unrealistically excessive.

To avoid any notable overstatement of Comeld-2 weight and cost effectiveness associated with the dimensional uncertainty, a conservative rough magnitude of the total length of the hybrid joint is utilized for cost assessment. According to the dimensions of the composite part of the deckhouse, which per LeGault's (2010) report is 39.6 m long by 18.3 m wide, the total length of the composite-metal seam L_j is presumed to be on the order of $L_j \sim 300$ m. This includes both outer seams at the metal skirts of the composite deckhouse and hangar shells as well as the metal-to-composite connections of the inner structural members of those topside structures.

For this conservative supposition, the direct weight savings are still significant. According to expression (4.28), they are $\Delta W \approx 40$ tons per ship, i.e., 3.7% of the weight of the entire composite topside structure. Per Lundquist's (2012) input, this is the sum of the weight of the composite part of the deckhouse $W_{DH} = 900$ tons and that of the hangar $W_H = 200$ tons, i.e., $W_\Sigma = 1100$ tons total.

The cost advantage ΔC of a Comeld-2 application is assessed using the following expression

$$\Delta C = (c_B - c_{C2})L_j - \frac{C_{n-r}}{N_s} \qquad (4.29)$$

where c_B is the cost rate of processing of a linear unit of the bolted joint; c_{C2} is the cost rate for Comeld-2 processing that comprises two major components, c_A for the adhesive bonding and c_{EB} representing the cost of EB metal surface protrusion, i.e.

$$c_{C2} = c_A + c_{EB} \qquad (4.30)$$

C_{n-r} in turn represents nonrecurring expenses, particularly associated with the capital cost of EB equipment and the cost of some additional engineering efforts to tweak a protrusion pattern corresponding to any imposed specific design requirements; N_s is the number of ships of the class for which the EB equipment is to be purchased, the protrusion process modified, and the related nonrecurring expenses amortized.

Per Brown's (2004) estimate, the processing cost of the bolted joint is $c_B = \$2190 \, m^{-1}$, versus that of the NJC adhesively bonded joint, $c_A = \$820 \, m^{-1}$. Apparently, this rate includes expenses covering the required machining of both the metal shoe and one side of the composite panel as well as the secondary bonding operation, embodying the NJC's main processing components attributable to the NJC joint (Simler and Brown, 2003).

In contrast, the Comeld-2 adhesive bond processing is associated with virtually no considerable extra cost as the molding of the composite part is co-processed with the metal parts adhesive bonding. Hence, it is fair to assume in the framework of the ROM estimate that the cost rate c_A for Comeld-2's adhesive bond component is roughly $c_A \approx 0$.

The EB processing cost component c_{EB} for the required protrusion of a 100-mm-wide metal surface should be about $\$190 \, m^{-1}$, as reported by Khodorkovsky et al. (2009), based on the courtesy input of *EBTEC Corporation.*

Given that two metal lap plates are required to implement Comeld-2 joining, the cost rate of EB metal treatment per unit length of the seam is $c_{EB} = \$380 \, m^{-1}$. Consequently, according to expression (4.30), the full cost rate of Comeld-2 processing is $c_{C2} \approx c_{EB} = \$380 \, m^{-1}$.

The capital cost of an EB station is $\sim\$235K$, for the capacity of a vacuum chamber sufficient to protrude the 100-mm-wide surface of 2.7-m-long plate, suitable for the intended protrusion of the metal lap plates. Another nonrecurring Comeld-2 cost for engineering of the EB protrusion is about $\$14.5K$ (Khodorkovsky et al., 2009). Jointly, the nonrecurring expenses amount to $C_{n-r} \approx \$250K$.

Presuming that, corresponding to the initial plans, a series of at least three *Zumwalt* class destroyers is to be built, i.e., $N_s = 3$, and the total joint length is to be at least $L_j \sim 300 \, m$, the direct cost savings ensuing from the conceived replacement of the bolted joint with Comeld-2 would be, according to expressions (4.29) and (4.30), as high as $\triangle C = \$460K$ per ship.

While this is a meaningful amount, the estimated direct cost savings is not the only source for the cost savings. That is augmented substantially by taking into account the weight saving of $\triangle W = 40$ tons per ship that is to be converted to cost savings. For instance, the high-performance CFRP currently employed for the *Zumwalt*'s topside structures could be partially replaced with a substantially less expensive but relatively weighty GFRP. Supposedly, the weight savings collected is sufficient to replace approximately 140 tons of CFRP. With such a material changeover, the indirect

Table 4.3 Anticipated ROM cost and weight savings

Joint type	Estimate		Conceivable savings			
	Weight, tons	Cost, $M	Direct Weight, tons	Cost, $M	Indirect Cost, $M	Total Cost, $M
Bolted joint	49	0.657				
Comeld-2	8.7	0.197	40	0.46	18	**18.46**

reduction of the ship construction cost on the order of ~$18M per ship might be added to the direct cost savings.

Table 4.3 summarizes the ROM assessment of the weight and cost savings per ship associated with the conceivable use of Comeld-2 to mount the composite topside structures on the metal base of the DDG-1000 *Zumwalt* class destroyers instead of the currently employed bolted joint.

It should be noted once again that while the presented savings expectations are significant, they are based on cost estimates for the current non-industrial, research stage of Comeld-2 development. Presumably, these savings would be notably increased as both major cost components of Comeld-2 implementation—the cost of EB Surfi-Sculpt material processing and that of nonrecurring engineering expenses—are reduced upon attaining broad industrial application of Comeld-2 technology.

4.10 COMELD-2 READINESS

The principal accomplishments of Comeld-2 development pertaining to the transition of this advantageous hybrid joining technology to fleet use are multifold. The US Navy technology readiness level (TRL) is constituted by the following:

- Validated feasibility and structural efficiency, TRL-3
- Developed principle manufacturing processes suitable for large hull construction application, TRL-4
- Selected effective math models and devised structural design analysis technique, TRL-4
- Demonstrated "in-fleet" reparability, TRL-4
- Selected optimized champion design configuration, TRL-4
- Fabricated full-scale champion Comeld-2 hybrid panels and test articles, TRL-4

- Executed comprehensive mechanical-environmental testing of Comeld-2 champion configuration, TRL-5
- Compiled experimental data sufficient to specify design allowables pertinent to hybrid hull outfitting with Comeld-2 material-transition/joint skirt, TRL-5.

On the whole, the attained results constitute a substantial basis for the targeted transition of the novel hybrid joining technology to fleet implementation. The upgraded Comeld-2 design along with the employed material processing techniques, both EB Surfi-Sculpt metal treatment and modified composite VIP, enable the sought improvement of structural efficiency of a hybrid joint and a hybrid hull as a whole. Structural performance of the selected Comeld-2 champion configuration is 2.5 times greater than that of its plain bond counterpart and demonstrates practically the same (5% higher) structural performance of a state-of-the-art bonded-bolted joint, the most structurally efficient, but excessively heavy, labor intensive, and costly joining option available to date.

The achieved level of Comeld-2 structural efficiency for short-term loading enables multiple functional, operational, weight, and cost benefits associated with realization of the hybrid hull concept. In view of this encouraging result, Comeld-2 is considered a sound alternative to the existing state-of-the-art hybrid joint options for assorted naval platforms utilizing the hybrid hull concept.

To authenticate this positive outlook, an expert's comparative scoring of the key performance parameters is presented in Table 4.4, which combines qualitative and available ROM quantitative data for both commonly used existing and novel joining options.

The following denotations identify the scored joining options.

(A)   Plain adhesive bonding
(B)   NJC adhesive bonding (Simler and Brown, 2003)
(C)   Ordinary fastened (bolted/riveted) joint
(D)   Combined bonded-fastened (bonded-bolted) joint
(E)   Bonded-pinned Comeld-2 joint.

The qualitative grades of the key performance parameters of distinct types of hybrid joints presented in Table 4.4 imply the following:

- Excellent—uncompromised performance associated with minimal risk of failure
- Fair—sufficient to satisfy the existing service requirements with lowered effectiveness and/or with moderate risk of malfunction

Table 4.4 Scoring matrix for hybrid joining options

Key performance parameters		Scores on joining options				
Categories	Sub-categories	A	B	C	D	E
Structural performance	Integrity w/hull structure					
	Reliability					
	Robustness comparing to "A" option	**1.0**	1.4[a]	2.4[b]	2.4	2.5
	Endurance					
	Impact resistance					
	Weight, kg/m	29	94[a]	162[a]	75[b]	29
	Watertightness					
Manufacturability	Repeatability					
	Labor intensity					
	Process automation					
Maintainability	Inspectability					
	Corrosion resistance					
	Reparability					
Cost	Processing, $/m	0[b]	820[a]	2190[a]	2190[b]	380
	Maintenance					
	Nonrecurring, $K/ship					83[c]
Associated risk	Readiness, TRL	8	6	8	6	5
	Risk mitigation					
Denotations		Excellent		Fair		Poor

[a] Per Brown's (2004) data.
[b] Best guess—relevant quantitative data are unavailable.
[c] For a three-ship series.

- Poor—incapable of providing requisite service performance, necessitating intensified labor operations, excessive material consumption and weight, and/or added specialized measures.

Although the given scoring is fairly subjective, it helps to provide an overview and comparison of assorted joining options with regard to all key

performance parameters, including structural performance, manufacturability, maintainability, cost, and capability of mitigating associated risk.

As can be seen, the Comeld-2 concept is a certain winner, earning the highest aggregated score, and due to this is the most appealing hybrid joining option for assorted heavy-duty applications. For this reason, Comeld-2 technology should be developed further, targeting a pilot and then a full-scale implementation for a naval platform utilizing the hybrid hull concept.

In particular, design, fabrication, and testing of a large full-scale technology demonstration panel (TDP) to attain TRL-6, required for implementation of a new technology for a pilot naval structure, should be fulfilled.

Additionally, serviceability characterization of hybrid structures needs certain methodological advancement, primarily with regard to possible thermomechanical interaction of dissimilar components during long-term service.

The existing ship design guidelines, including the state-of-the-art Design and Classification Rules/Guides (2003, 2012), are mainly dedicated to a separate application of either metal or composite structures and do not specifically address peculiarities inherent to service behavior of hybrid, heterogeneous structures. Hence, an advanced analytical technique is necessary to bridge the methodological gap. An advanced concept of serviceability characterization of composite and hybrid structures undergoing changing force-ambient operational exposure is delineated in Chapter 5 in an attempt to meet this requirement.

REFERENCES

Anon., 2011. DDG 1000 Deckhouse Base Joint. NMC – the Naval Metalworking Center. Available from www.ncemt.ctc.com/index.cfm?fuseaction=projects.details&projectID=233.

Anon., n.d. Damage Control Skills and Tricks of the Trade 020: Advanced Damage Control. NAVSEA System Command. Available from http://www.dcfpnavymil.org/Library/dctricks.htm.

ASTM D3039/D3039M-00, 2000. Standard Test Method for Tensile Properties of Polymer Matrix Composite Materials.

ASTM D5229/D5229M-92, 1992. Standard Test Method for Moisture Absorption Properties and Equilibrium Conditioning of Polymer Matrix Composite Materials.

Boyd, S.W., Blake, J.I.R., Shenoi, R.A., Kapadia, A., 2004. Integrity of Hybrid Steel-to-Composite Joints for Marine Application. Proceedings of the Institution of Mechanical Engineers, Part M: Journal of Engineering for the Maritime Environment, 235–246.

Brown, L., 2004. Composite to Steel Joints – Developed for the Next Generation Surface Combatant. Technical Presentation, the ASM International Indianapolis Chapter.

Dance, B.G.I., Kellar, C., 2010. Workpiece Structure Modification. Patent 7667158 B2, USA. Available from http://www.google.co.ck/patents/US7667158.

Khodorkovsky, Y., Shkolnikov, V., 2010. Composite-to-metal joining technology. Final Report No. 10-05, STTR Phase II Base, Contract: N00014-09-C-0331, Beltran, Inc., Brooklyn, NY, 87pp.

Khodorkovsky, Y., Shkolnikov, V., 2011. Composite-to-metal joining technology. Final Report No. 11-07, STTR Phase II Option 1, Contract: N00014-09-C-0331, Beltran, Inc., Brooklyn, NY, 55pp.

Khodorkovsky, Y., Mouring, S., Shkolnikov, V., 2009. Advanced hybrid joining technology. In: Proceedings of the 1st International Conference on Light Weight Marine Structures, Glasgow, UK, September 7–8, 10pp.

LeGault, M.R., 2010. DDG-1000 Zumwalt: stealth warship – U.S. navy navigates radar transparency, cost and weight challenges with composite superstructure design. Composites Technology, February. Available from http://www.compositesworld.com/articles/ddg-1000-zumwalt-stealth-warship.

Loup, D., 2010. Request for Release of the AHMST's US Elemental Hybrid Joints Test Results (Task 7) to ONR STTR Contractor: Beltran, C CIV NSWCCD W. Bethesda, 6550, douglas.loup@navy.mil, October 28.

Lundquist, E., 2012. US Navy: DDG 1000's composite deckhouse milestone. Maritime Reporter & Engineering News 26–28, January.

Miyano, Y., Nakada, M., Sekine, N., 2005. Accelerated Testing for Long-Term Durability of FRP Laminates for Marine Use, Journal of Composite Materials, 39 (1), 5–20.

Mouring, S.E., 2009. Performance of hybrid metal-to-composite joints, the U.S. Naval Academy. Report N00014-09-WR-2-0159.

Mouring, S.E., 2010. Tensile Test Results of the Repaired Joint Articles. E-mail mouring@usna.edu, February 12.

NAVEDTRA 43119-J, 2008. Personnel Qualification Standard for Damage Control (DC). Naval Personnel Development Command, July, 294pp. Available from http://www.dcfpnavymil.org/Library/dcpubs/43119-J%20Damage%20Control%20(DC)%20[1].pdf.

Osborne, T., 2014. An introduction to resin infusion. Reinforced Plastics 25–29, January/February.

Regel, V.R., Slutsker, A.I., Tomashevsky, E.E., 1972. The Kinetic Nature of the Strength of Solids, Progress of Physical Sciences, 106 (2), Moscow: Russia, 193–228 (Кинетическая Природа Прочности Твёрдых Тел).

Roark, R.J., Young, W.C., 1975. Formulas for Stress and Strain, fifth ed. McGraw-Hill Book Company, NY, 624pp.

Shkolnikov, V.M., 1995. Principles of Computational Life Prediction of Composite Structures. In: Questions of Material Science, 3, Russia, SPb, CSRI CM 'Prometey', pp. 30–38 (Принципы Расчётного Прогнозирования Прочности Корпусных Конструкций из Полимерных Композитов в Аспекте Практического Приложения).

Shkolnikov, V.M., 2013. Material-Transition Structural Component for Producing of Hybrid Ship Hulls, Ship Hulls Containing the Same, and Method of Manufacturing the Same. Patent 8,430,046 B2, USA.

Shkolnikov, V.M., Khodorkovsky, Y., 2011. To-date advancement of bonded-pinned composite-to-metal joining technology. In: Proceedings of SAMPE-2011 Conference, Long Beach, CA, May 23–26, 13pp.

Shkolnikov, V.M., Dance, B.G.I., Hostetter, G.J., McNamara, D.K., Pickens, J.R., Turcheck Jr., S.P., 2009. Advanced hybrid joining technology-OMAE2009-79769. In: Proceedings of the ASME 28th International Conference on Ocean, Offshore & Arctic Engineering, OMAE2009, Honolulu, Hawaii, May 31–June 58pp.

Simler, J., Brown, L., 2003. 21st century surface combatants require improved composite-to-steel adhesive bonds. The AMPTIAC Quarterly 7 (3), 21–25. Available from, http://ammtiac.alionscience.com/pdf/AMPQ7_3ART03.pdf.

CHAPTER 5

Serviceability Characterization

Being in service, a ship's hull undergoes broad-ranging force-ambient exposures, the deteriorative influence of which greatly affects structural behavior, interaction, and load-bearing capability of hull structures. This relates to every loading occurrence, be that impulse, vibration, short-term quasi-static, or long-term protracted or repetitive force exposure. Accurate characterization and prediction of service performance embody, in fact, one of the principal problematic aspects of structural engineering, which is particularly challenging regarding heterogeneous structural systems. Chapter 5 addresses methodological aspects of serviceability evaluation of composite and hybrid structures undergoing changing operational loading. The imparted analytical approach represents an attempt to advance the accuracy of serviceability evaluation based on the kinetic theory of fracture and other well-justified physical and math models.

5.1 EXISTING APPROACH

Whatever the particular destination of an engineered structure, it needs design verification of requisite robustness under anticipated force-ambient operational exposures for the assigned lifetime. While in service, a ship's hull undergoes broad-ranging force-ambient exposures which greatly affect structural behavior, interaction, and load-bearing capability. This relates to every loading occurrence, be it impulse, vibration, or short-term quasi-static, or long-term protracted or repetitive force exposure. Most of these are typically accompanied by altering temperature and/or other deteriorative factors of the harsh operational environment.

An accurate characterization and prediction of service performance is one of the principal problematic aspects of structural engineering. It is particularly challenging for heterogeneous structural systems due to the possibly substantial diverse service properties of the utilized materials and, in the case of a PMC application, enhanced sensitivity to force-ambient conditions.

Along with employment of advanced math models, computational algorithms, and computer software for design stress/strain analyses, serviceability evaluation demands extensive experimental verification of analytical

estimates for all PMC material systems chosen for a particular hull structure. Massive static, dynamic, and ambient-mechanical fatigue testing of material coupons, full-scale test articles, and prototype structures is routinely required. While these tests are typically quite time consuming and costly, shortage of those is fraught with risking either unwarranted excessiveness or deficiency of structural robustness. Either of such outcomes is objectionable, as the first may cause extra weight, decreased structural efficiency, and excessive cost; whereas the other would downgrade serviceability of the designed structure, compromising the reliability and safety of its operation.

The heterogeneity of hybrids aggravates the problem due to the diversity of structural and physical properties when utilizing dissimilar material systems which may cause considerable extra stressing, affecting the load-bearing capability of the entire hybrid hull or shortening its lifetime.

> The heterogeneity of hybrids aggravates the problem of accurate serviceability characterization due to the diversity of structural and physical properties of utilized dissimilar material systems that may cause considerable extra stressing, affecting load-bearing capability of the entire hybrid hull or shortening its lifetime.

The differences in densities, moduli of elasticity, fatigue performance, and rates of thermal expansion are the main potential contributors pertaining to this phenomenon.

For this reason, direct translation of existing standards specifying design allowances for mono-material structures may be deficient for a hybrid structure, leading to either under- or over-design, possibly succeeding with noticeable cutback of anticipated structural and other functional benefits. Therefore, it is vital to enhance the existing strength reconciliation routine to properly reflect the peculiarities intrinsic to hybrid structures and their service.

The common approach for verification of a structure's robustness for a given variety of operational loadings and an assortment of employed material and structural systems is design reconciliation of the maximum combined stress σ_m (and/or deformations ε_m), induced within the structure with design–allowable stress σ_a (deformation ε_a). Analytically, this is expressed by strength acceptance criteria to be satisfied:

$$\sigma_m \leq \sigma_a \tag{5.1}$$

$$\varepsilon_m \leq \varepsilon_a \tag{5.1a}$$

Specification of the design allowables σ_a (ε_a) requires diverse considerations. They are determined based on the mechanical and physical properties
of structural materials and downgraded corresponding to anticipated deterioration of those materials under long-term force-ambient operational exposure and the given safety margin.

PMCs are more sensitive to the deteriorative influence of the harsh
marine operational environment than are structural metals. Each partial
force-ambient exposure contributes to deterioration of the PMC in service.
For this reason, the whole array of deteriorative factors of ship operation
should be taken into account to ensure proper robustness of the compos-
ite/hybrid structure for its service life.

> Due to enhanced sensitivity of PMCs to the deteriorative influence of the harsh
> marine operational environment, essentially the whole array of deteriorative
> factors of ship operation should be taken into account to ensure proper
> robustness of the composite/hybrid structure for its service life.

In general, those factors consist of length of sustained or repetitive loading, thermal exposure, seawater moistening, and exposure to UV and/or
other irradiation.

The safety margin in turn is a measure of structure criticality for proper
functioning of the ship and its components, uncertainties pertinent to characterization of the anticipated operational loading, accuracy of employed
math models used for stress/strain and failure analyses, and possible scattering
of PMC properties due to the variable properties of unprocessed PMC constituents and versatility of applied material processing during composite/
hybrid structure manufacturing.

Accordingly, the design stress allowables σ_a are expressed as

$$\sigma_a = \frac{\sigma_d}{f_s} \tag{5.2}$$

where σ_d is a design stress value that implies PMC ultimate strength S,
reduced corresponding to the anticipated material deterioration during its
service; f_s is safety factor that quantifies the assigned safety margin.

It should be noted that the safety factor may be alternatively presented
within the maximum stress value σ_m of Equation (5.1), as it is the practice
in other industries, e.g., airframes design.

To quantify PMC deterioration during its service within a structure, partial knock-down coefficients k_i, corresponding to the intensity and length of operational exposures and resistance to those of a particular material system, are typically applied. The combined deteriorative effect of diverse deteriorative factors as a product is the most common approach, expressed by

$$\sigma_d = S\prod_{i=1}^{m} k_i \qquad (5.3)$$

where S is the ultimate strength of PMC for its particular failure mode determined under normal testing conditions; m is the number of deteriorative factors to be taken into account.

Note that the two expressions (5.2) and (5.3) are often combined into one, proceeding with a generalized safety factor $\hat{f}_s$. In this case, to represent all strength-reducing factors counted in (5.2) and (5.3), the generalized safety factor $\hat{f}_s$ is expressed, as in the references (DNV, 2012; Greene, 1997; RD 5.1186-90; Smirnova et al., 1984),

$$\hat{f}_s = f_s\left(\prod_{i=1}^{m} k_i\right)^{-1} \qquad (5.4)$$

or

$$\hat{f}_s = \frac{1}{\prod_{i=1}^{n} k_i} \qquad (5.5)$$

Here n is the extended number of partial factors taken into account, including both groups of reduction coefficients, reflecting material deterioration for long-term operation and the quantified safety margin.

This approach makes determination of design allowables for a composite/hybrid structure similar to that for common metal structure design, expressed as

$$\sigma_a = \frac{S}{\hat{f}_s} \qquad (5.6)$$

where, in the case of a metal structure, S would stand for the yield point S_y.

Evidently, both these expedients produce the same values for design allowables. Nevertheless, the discrete representation of the two distinct

groups of reduction factors is preferable, as it precludes the designer experiencing the illusion that the safety margin is excessive.

Currently, determination of the partial knockdown coefficient k_i heavily relies on empirical relations derived from extensive experimentally acquired data on PMC serviceability under assorted force-ambient exposures.

The knockdown factors may be determined with different levels of detail, employing distinct computing procedures. In fact those provided by assorted design guidelines vary significantly. For instance, ABS (2007) and DNV (2012), both relevant to essentially the same class of high-speed naval craft, express similar, quite simplified requirements for knockdown determination. In particular, for the primary PMC structural components of those naval crafts, including the bottom shell, side shell, decks, superstructure, and deckhouses, the allowed stress is the same, $\sigma_a = 0.33S$, implying a generalized safety factor of $\hat{f}_s = 3.0$.

Contrary to this, DNV (2010) that represents another worthy source of the design guidelines pertaining to heavy-duty marine composite structures, particularly to structural components of offshore platforms, provides meticulously detailed and distinguished values for design allowances consistent with the outcome of a target study reported by Echtermeyer et al. (2002).

A compromise between these two extreme approaches is given by Smirnova et al. (1984). This provides an array of tabulated partial coefficients k_i for expression (5.5) with reasonable detail, for anticipated ranges of operational conditions, criticality of a particular structure, and other factors influencing design considerations.

Table 5.1 replicates the generalized array of downgrading coefficients k_i, originally presented by Smirnova et al. (1984), with slightly tweaked values corresponding to accuracy of the currently available computing means, properties of modern PMC compositions, and the superiority of up-to-date manufacturing technologies.

Note that data imparted in Table 5.1 do not precisely define the particular coefficients k_i. Rather, they demonstrate a variety of magnitudes for the discrete knockdown/safety factors and estimate both the degree of material deterioration and value of the generalized safety factor for a particular structure operational scenario.

Table 5.1 Performance downgrading factors

i	Category	Description	Coefficient values
1	Criticality of structure	Vital for ship serviceability	0.85–0.90
		Significant for ship operation/ crew functioning	0.90–0.95
		Enabling ancillary equipment and habitability	0.95–1.00
2	Characteristic nature of operational loading	Stochastic	0.85–0.95
		Deterministic	0.90–1.00
3	Accuracy of computer models and algorithms	Simplified math models	0.90–0.95
		2D, 3D FE models	0.95–1.00
4	Strength reconciliation criteria	Linear failure criteria	0.90–0.95
		2D, 3D failure criteria	0.95–1.00
5	Intricacy of structure shape/processing	Thin panels w/insignificant curvature	0.95–1.00
		Thick/variably-thick panels w/limited curvature	0.90–0.95
		Complex structures/joints	0.85–0.90
6	QA of material processing	All steps control	0.95–1.00
		Major steps control	0.90–0.95
7	Moistening/aggressive environmental exposure (for unprotected structures)	Sustained	0.85–0.90
		Repetitive	0.90–0.95
		Sporadic to none	0.95–1.00
8	Ambient temperature	Up to 80% of heat distortion temperature	0.85–0.90
		Up to 50% of heat distortion temperature	0.90–0.95
		Normal (room) temperature	1.00
9	UV/other irradiation (for unprotected structures)	Regular	0.85–0.90
		Sporadic to none	0.95–1.00
10	Temporal category of loading	Sustained ($\tau \geq 10^4$ hours)	0.30–0.60
		Low cyclic	0.45–0.75
		Short term	0.95–1.00

As can be found based on Equation (5.5) and data provided in Table 5.1, the generalized safety factor ranges widely, roughly as $1.0 \leq \hat{f}_s \leq 5.0$, depending upon the original structural parameters, anticipated operational conditions, and the aftermath severity in the event of structure failure. For a particular material system and its intended structural application, the given ballpark values should be refined based on pertinent experimental data.

It should be accentuated that moderate detail of the design down-grading factors, such as that provided by Smirnova et al. (1984), allows for justifiable optimization of a composite/hybrid hull structure, while not dramatically complicating the design strength reconciliation routine. This is probably the main reason for keeping this approach in effect, till now reflected in the current design guidelines (RD 5.1186-90).

5.2 PREREQUISITES OF METHODOLOGY ADVANCEMENT

Although the existing procedures for specification of design allowables have been used for decades, enabling reliable robust design of assorted naval plat-forms, they are not sufficiently expedient for up-to-date applications. The methodological inadequacy primarily relates to the empirical basis of the existing approach, which necessitates execution of extensive experimental programs to engender property data sufficient to support specification of design knockdowns on multiple material compositions and anticipated force-ambient operational exposures.

The material variety inherent to PMCs that essentially reflects the down-side of the advantageous diversity of structural properties of a composite largely depends on innate traits, which include:

- Types of PMC main constituents, both fiber reinforcement and polymer resin
- Forms of fiber material, e.g., woven fabric; non-crimp stitched fabric; chopped mat; and mat veil
- Fiber material layup and its processing, e.g., fabric hand layup; tow place-ment; and automated tape layup
- Molding technology, i.e., open mold; closed-mold VIP; and prepreg utilization.

PMC variety unavoidably leads to expansion of relevant experimental pro-grams, leading to excessive time consumption and monetary expenditure which may together substantially delay and increase the cost of acquisition activities. Often this is a prime factor inhibiting use of the new material systems.

Due to this, it may become impractical to follow the course of a direct experiential determination of structural performance for a whole range of candidate structural materials and structural components with regard to each and every anticipated loading exposure, despite the great dependability of this approach.

It may become impractical to follow the course of a direct experiential determination of structural performance for a whole range of candidate structural materials and structural components with regard to each and every anticipated loading exposure, despite the great dependability of this approach.

There is an issue that further compromises the currently employed method. It relates to inadequate interpretation of property data ensuing from performed experimental programs that may occur as actual parameters of performed testing are overlooked.

As asserted some time ago (Michaelov et al., 1997; Regel et al., 1972), a moderate variation of force-ambient parameters is often neglected in analysis of PMC test data. This practice probably originates from the legacy of metals testing, for which such neglect is not as meaningful as it can be for structural PMCs.

Whatever the reason, the disregard for a subtle variation of loading parameters may result in a perceptible misrepresentation of PMC properties data, potentially impairing an undertaken structural design.

The disregard to a subtle variation of loading parameters may result in a perceptible misrepresentation of PMC property data, potentially impairing an undertaken structural design.

Inadequate interpretation usually relates to mixed utilization of two distinct sets of test data, one ensuing from a monotonic short-term loading and the other gained from fatigue testing. While this is a viable approach as the difference in loading rates is counted, it may bring a deceptive result when that difference is ignored.

According to test standards (e.g., ASTM D 3039/D 3039M-00), the typical required length of a short-term test loading ϑ to produce specimen failure is within 1 to 10 min, i.e., $60 \leq \vartheta \leq 600$ s. The frequency of a cyclic loading f in turn is expected to be within a range of $10 \leq f \leq 30$ Hz, which usually corresponds to the requirement to keep it low enough to avoid significant temperature variations (ASTM D 3479/D 3479M-96). The encountered difference in roughly three or more orders of magnitude of the loading rate is sufficient to cause substantial deviation of the stress-durability $(S - N)$ diagram from its true view—typically linear, with semi-logarithmic axes—which is fraught with distortion of the factual test data.

Contrary to this, properly unified test data would allow for correct specification of the design allowables. This is especially important for large composite and/or hybrid structures, for which even a minor mismatch between anticipated and actual material performance could significantly increase the weight and cost effectiveness of ship hull construction.

The extensiveness of test programs as well as the database size issue can now be addressed with utilization of analytical methods incorporating advanced material failure and damage models which have shown promise for use in composite material property and serviceability prediction. Particularly, this relates to characterization of PMC deterioration under long-term force-ambient exposure, which represents a principal contributing factor to structural performance downgrade. Hence, trustworthy analytical models capable of adequately reflecting a material's response to a long-term loading allied with its alternation are necessary to allow for meaningful truncation of the obligatory experimental programs.

An analytical approach introduced in the references (Shkolnikov, 1995, 2007) is capable of meeting this demand with the accuracy suitable for ship hull design. The following represents prerequisites and rationales of this methodological advancement.

First, two major groups of the force-ambient factors of operational exposure experienced by a ship hull in service should be differentiated by the way they influence a material's service performance. One group comprises environmental impacts, such as seawater moistening or UV irradiation, which mainly affect one or two external plies of a composite layup. The other unites most force and thermal exposures affecting the entire PMC layup.

Following this distinction, deterioration of the external and internal plies of a PMC layup—especially for the thick-walled structures typical for a ship hull application—may be addressed differently. For external plies, the methodology should provided protective measures, such as enrichment of the resin content, use of nonstructural protective coating, and/or an application of an extra "sacrificial" layer of the utilized structural PMC.

Inasmuch as the outmost external layer(s) is mainly to protect the primary laminate's part, it can be omitted from the design strength reconciliation, with its focus on the primary part and its serviceability. Being protected from the direct environmental impact, the primary part of the composite laminate is mainly subjected to just two determinant factors of operational exposure—the force and the ambient temperature. Accordingly, serviceability of a PMC subjected to assorted operational exposures may be characterized by the influence of only those two determinants.

As is known, the endurance of most structural PMCs under constant stress σ is congruent with the empirical expression

$$\tau = A\exp(-\alpha\sigma) \qquad (5.7)$$

where parameters A and α represent time-dependency characteristics of a PMC, invariant to the stress level σ. They are, however, sensitive to stress alteration as well as to changing of ambient temperature, if imposed. Accordingly, the magnitude of A and α would alter, pertaining to deterioration of the PMC under changing force-ambient exposures.

Meanwhile, the kinetic theory of fracture (Regel et al., 1972; Zhurkov, 1984; Zhurkov and Tomashevsky, 1966) provides a dependable foundation for an analytical representation of PMC endurance based on the kinetic characterization of the fracture of solids. For the kinetic-based concept, a material's fracture is considered as a time process, the rate of which is determined by mechanical stress and temperature. Specifically, this implies that the kinetic thermo-fluctuation controls the origin and growth of cracks of all levels, from incipient submicroscopic cracks with sizes of tens of angstroms to macroscopic cracks.

As stated by Zhurkov (1984) and Regel et al. (1972), the crack growth under loading has been studied for more than 50 assorted solid materials, including structural PMCs. Microfilm, electron microscopes, and electron paramagnetic resonance have all been employed to measure long-term crack growth, in some cases for several decades. Those measurements have shown that the growth of main cracks occurs in three phases:

(1) Origin of incipient microcracks as a result of breaching inter-element bonds
(2) Development and accumulation of these microcracks to critical concentration
(3) Coalescence of secondary microcracks which arise at the tip of the main one.

The interaction and growth of cracks of all dimensions, from submicroscopic and up to macroscopic, occur at an exponential rate of the applied mechanical stress and the reciprocal test temperature. The principal outcome of this meticulous experimental investigation is validation of the initial assumption that the process of material damage under applied loading corresponds well to the kinetic notion of fracture of solids. Without loading, the probability of overcoming potential barriers is low and motions of kinetic microelements are equally probable. In other words, rare breakage of inter-element bonds is compensated for by their restoration. The application

of a force upsets this balance, changing distances between the microelements and weakening their links along the direction of the force.

The results of the experimental examination of material endurance suggest that a universal relation between lifetime τ, mechanical stress, σ, and absolute temperature T exists and can be written in the form of a kinetic operation, as

$$\tau = \tau_0 \exp\frac{U_0 - \gamma\sigma}{kT} \tag{5.8}$$

where k is the Boltzmann constant and τ_0, U_0, and γ are constant coefficients characterizing the time dependency of a material's load-bearing capability.

It has been found that parameter τ_0 is essentially the reciprocal of the natural oscillation frequency of atoms in a solid. The constant U_0 represents the binding energy on the atom scale for assorted solids, including structural metals and PMCs. The coefficient γ is proportional to the disorientation of the molecular structure which predetermines its wide variability upon constitution of a particular material.

Expression (5.8) has a typical thermo-fluctuation view, as multiplier $\exp\left(-\frac{U_0}{kT}\right)$, or its converted quantity—the "Boltzmann factor"—is widely used to describe various thermo-fluctuation processes, such as chemical reactions, diffusion, and vaporization, among others. Note that while the reported experiments were mainly executed at material tension, it was ascertained that other loading cases manifested similar relations between time, stress, and temperature, as expressed by relations (5.7) and (5.8)

With regard to an engineering application, it is preferable to express relation (5.8) with the equivalent expression

$$\tau = \tau_0 \exp\left(\frac{\varpi_0 - \alpha_0\widetilde{\sigma}}{\widetilde{T}}\right) \tag{5.9}$$

which implies:

$$\widetilde{T} = \frac{T}{T_s} \tag{5.10}$$

$$\overline{\omega}_0 = \frac{U_0}{kT_s} \tag{5.11}$$

$$\alpha_0 \equiv \alpha S = \frac{\gamma S}{kT_s} \tag{5.12}$$

$$\widetilde{\sigma} = \frac{\sigma}{S} \tag{5.13}$$

Here T_s stands for standardized normal (room) temperature on the Kelvin scale, i.e., $T_s = 293.15\,°K$; S is the ultimate strength of a PMC defined at room temperature and other standardized testing conditions.

Apparently, as ambient temperature is equal to the standard, i.e., $T = T_s$, expression (5.9) adopts its simplified form

$$\tau = \tau_0 \exp(\varpi_0 - \alpha_0 \widetilde{\sigma}) \qquad (5.14)$$

which is completely congruent with the conventional empiric relation (5.7), as the parameters thereof are $A = \tau_0 \exp(\varpi_0)$; $\alpha = \frac{\alpha_0}{S}$.

However, contrary to relation (5.7), expression (5.9) enables characterization of the length of service life of a composite structure, taking into account both principal deteriorative factors, mechanical stressing and temperature exposure.

Apparently, expression (5.9) can be rewritten in the conventional form relative to stress rupture

$$\widetilde{\sigma} = B_{\tau,T} - \beta_{\tau,T} \log_{10}(\tau) \qquad (5.15)$$

And, similar to (5.9), relation (5.15) reflects the influence of ambient temperature along with the accustomed characterization of load-bearing capability as a function of the length of loading.

The values of parameters $B_{\tau,T}$ and $\beta_{\tau,T}$ according to (5.9) are

$$B_{\tau,T} = \frac{\varpi_0 + \widetilde{T} \ln(10) \log_{10}(\tau_0)}{\alpha_0} \qquad (5.16)$$

$$\beta_{\tau,T} = \frac{\ln(10)}{\alpha_0} \widetilde{T} \qquad (5.17)$$

Note that because the dimensionless parameters $\alpha_0; \varpi_0; \tau_0$ are invariant to direction of applied loading, expression (5.9) is valid for characterization of PMC performance regardless of the orientation of its orthotropy axes.

> Imparted kinetic-based analytical expressions are valid for characterization of a PMC's endurance regardless of the orientation of its orthotropy axes.

Note also that although parameter $\widetilde{T}$ nominally reflects the ambient temperature alteration, it may rather represent any environmental factor and/or a combination of factors when it is a preferable option to the consideration of environmental factors suggested above.

Further, as far as Equation (5.9) presumes unchanging force-temperature loading conditions, to characterize PMC performance under altering

loading, Equation (5.9) needs be paired with a trustworthy hypothesis on material damage accumulation. Bailey's integral generalizing the hypothesis on linear damage accumulation extends such an opportunity to material deterioration (damage fraction), D being acquired as a result of an applied arbitrary loading.

Regarding alteration of both principal loading components, mechanical stress $\sigma(t)$ and temperature $T(t)$, Bailey's integral is expressed as

$$D = \int_0^{\vartheta} \frac{dt}{\tau(\sigma(t), T(t))} \qquad (5.18)$$

It presumes that damage fraction D acquired during ϑ length of loading may be within a range of $0 \leq D \leq D_F$, where value $D_F = 1$ implies the ultimate damage fraction to be attained at the ending failure point of material lifetime ϑ_F.

While providing a favorable opportunity for evaluation and prediction of material serviceability under arbitrary loading, a practical application of the linear damage accumulation rule has not gained common approval from the mechanical engineering community to date. Mainly, the bias against it is due to considerable mismatch between predicted and actual experimental data accompanying validity verification of this analytical technique. Meanwhile, the issue seems to relate not so much to the linear concept itself as to a misinterpretation of the experimental data—an issue that needs to be specifically addressed.

As is well known, the linear damage accumulation rule was introduced by Palmgren (1924) and further developed by Miner (1945). In Palmgren's version, a damage fraction D_i, accumulated under constant stress level $\tilde{\sigma}_i$ during length of loading ϑ_i, is equal to a ratio of this length of loading ϑ_i to the lifetime τ_i under the same stress level σ_i, i.e.,

$$D_i = \frac{\vartheta_i}{\tau_i} \qquad (5.19)$$

In Miner's version, the damage fraction D_i, accumulated during loading cycles at constant stress amplitude σ_i, is equal to the ratio of the applied cycles count n_i at this stress amplitude σ_i to the fatigue life N_i under this stress amplitude σ_i, i.e.,

$$D_i = \frac{n_i}{N_i} \qquad (5.20)$$

Similar to Bailey's integral (5.18), failure of either Miner's or Palmgren's version occurs when the ultimate accumulated damage $D_F = 1$, i.e., when

$$D_F \equiv \sum_{i=1}^{N} D_i = 1 \qquad (5.21)$$

Substantial test data have been generated in an attempt to verify validity of the linear rule for practical use. In most test cases, a two-round loading is employed, with specimen testing at an initial stress level σ_1 for a certain number of cycles n_1 followed by a second stress level σ_2 for n_2 cycles, until failure ensues.

Results of the original tests, performed by Miner (1945), showed that the actual ultimate damage fraction considerably deviated from its nominal value $D_F = 1$, in the range of $0.61 \leq D_F \leq 1.45$. Other researchers have shown a substantially larger D_F variation, up to $0.18 \leq D_F \leq 23.0$.

Along with the observed discrepancy in magnitude of the ultimate damage fraction, experimental verification of the concept also tends to manifest different outcomes depending upon the sequence of stress application, high-to-low ($\sigma_1 > \sigma_2$) versus low-to-high ($\sigma_1 < \sigma_2$) stress, which is not reflected in the theory.

Apparently, the loading parameters should be selected to get specimen failure at the second round of a two-round trial, no matter the loading sequence. Also, the total length of the two-round trials should be anticipated to be about equal, with the first round of the second trial as long as the second round of the first trial and vice versa. Otherwise, the total lengths of the two diverse trials would not be alike, but instead be substantially different.

In reality, the high-to-low test is typically significantly longer than the low-to-high test. The noted contradictions of significant variation of the ultimate damage fraction value and dependency of endurance upon loading sequence are usually asserted by examiners of the linear damage accumulation concept as a persuasive argument against its validity. However, the results of the above experiments are actually quite encouraging and do testify to concept validity.

Applying the linear damage accumulation rule it should be realized that lifetime expressed either as $\tau(\tilde{\sigma})$ or its analogue $N(\tilde{\sigma})$ is an exponential function. This implies that only negligible scattering of strength properties of tested specimens can produce an ultimate damage fraction value nearing $D_F = 1$. However, in reality, the result will range widely, corresponding to actual deviation of strength properties from their nominal values.

For instance, if we realistically assume that time-dependency parameters of PMC relevant to expression (5.15) are $\beta_{\tau,T} = 0.09$ and $B_{\tau,T} = 1.034$ and the relative stress being induced is $\tilde{\sigma} = 0.75$, the nominal endurance under this stress level, according to expression (5.15), would be $\tau = 1396$ s.

As the ultimate strength of test specimens of the same batch varies within a reasonable range with a coefficient of variation $CV = \pm 5\%$, the anticipated endurance of individual specimens of that batch would vary within a range $510 \leq \tau \leq 3470$ s. The corresponding range of the ultimate damage fraction would be $0.37 \leq D_F \leq 2.49$, which is a good match to the above experimentally acquired range, asserted as excessive. If the minor temperature variation that usually accompanies long-term testing is also taken into account, the resulting range of the ultimate damage fraction increases further.

The issue of loading sequence has a similar straightforward explanation. If we further assume that two two-round trials are executed, then according to expressions (5.19) and (5.21) the anticipated ultimate damage fraction $D_F = 1$ should satisfy the following relations

$$D_F \equiv \frac{\vartheta_{1,1}}{\tau_{1,1}} + \frac{\vartheta_{1,2}}{\tau_{1,2}} = 1 \qquad (5.22)$$

$$D_F \equiv \frac{\vartheta_{2,1}}{\tau_{2,1}} + \frac{\vartheta_{2,2}}{\tau_{2,2}} = 1 \qquad (5.23)$$

or relative to lengths of the second loading rounds

$$\vartheta_{1,2} = \left(1 - \frac{\vartheta_{1,1}}{\tau_{1,1}}\right)\tau_{1,2} \qquad (5.24)$$

$$\vartheta_{2,2} = \left(1 - \frac{\vartheta_{2,1}}{\tau_{2,1}}\right)\tau_{2,2} \qquad (5.25)$$

where the first subscript indicates the trial number and the second stands for the round number.

Let us also assume that the first round of each of the two two-round trials is to be maintained for $\vartheta_{1,1} = \vartheta_{2,1} = 400$ s, whereas the second rounds $\vartheta_{1,2}, \vartheta_{2,2}$ are to last till specimen failure. Stress levels for those two trials are as follows.
(1) For the first round of the first trial, the relative stress is $\tilde{\sigma}_{1,1} = 0.55$; for the second round of the first trial, it is $\tilde{\sigma}_{1,2} = 0.75$.
(2) For the first round of the second trial, the relative stress is $\tilde{\sigma}_{2,1} = 0.75$; for the second round of the second trial, it is $\tilde{\sigma}_{2,2} = 0.55$.

Then, for the time-dependency parameters used above ($\beta_{\tau,T}=0.09$; $B_{\tau,T}=1.034$), we should anticipate endurances of $\tau_{1,1}\equiv\tau_{2,2}=228{,}980$ s and $\tau_{1,2}\equiv\tau_{2,1}=1396$ s.

With substitution of these endurance anticipation values in expressions (5.24) and (5.25), we can expect the length of the second rounds to be:

$$\vartheta_{1,2}=\left(1-\frac{400}{228{,}980}\right)1396=1394\,\text{s}$$

$$\vartheta_{2,2}=\left(1-\frac{400}{1396}\right)228{,}980=163{,}370\,\text{s}$$

These numbers imply that the anticipated total length of the first trial is to be $\tau_1=\vartheta_{1,1}+\vartheta_{1,2}=1794$ s; and that of the second trial is to be $\tau_2=\vartheta_{2,1}+\vartheta_{2,2}=163{,}770$ s.

Taking into account a possible variation of the ultimate strength $CV=\pm5\%$, the total lengths of the two-round test loading will further expand the difference. For the first trial it would be in the range of $908\leq\tau_1\leq3870$ s; for the second trial it would be $24{,}000\leq\tau_2\leq395{,}500$ s—the result that speaks for itself.

It is appropriate to mention that obtaining such a difference from a real experiment is sometimes interpreted as an adverse effect of an initially high stress application in the first case, or a favorable result due to preceding material "training" of a moderate initial stress application in the second case.

It should be emphasized that in contrast with the significant scattering of endurance, the inverse operations regarding determination of either damage fraction or ultimate stress produce a quite stable result due to the same exponential nature of the utilized math model.

Overall, the imparted rationales provide convincing reasons for the complete consistency of the linear damage accumulation rule with the experimental data. This indicates favorable opportunity for its use for characterization of PMC serviceability and specification of design allowables for a composite/hybrid structure.

Owing to complete the consistency of the linear damage accumulation rule with experimentally gained data, it is favorable to use it for characterization of PMC serviceability and specification of design allowables for a composite/hybrid structure.

It is particularly worthwhile with the use of Bailey's integral (5.18), which allows for evaluation of structural performance of a composite

undergoing an arbitrary loading profile over its service life. It is also instrumental for verification of that composite's performance compared to structural performance of the metal counterpart of the hybrid structure, enabling quantifiable monitoring of possible migration of the weakest link over the structure's long-term service.

5.3 SERVICEABILITY AT CONVENTIONAL LOAD CASES

For most practically significant load cases integral (5.18) needs to be solved numerically. Nevertheless, if the ambient temperature does not change, a few closed-form solutions are available for the simplest loading profiles, such as monotonic constant-rate loading and pulse cycling with standard broken-line or haversine waveforms. The following illustrates the applicability of the introduced analytical approach for determining the significance of loading parameters on PMC structural performance.

5.3.1 Monotonic constant-rate loading

With regard to monotonic constant-rate loading to failure at the point of time ϑ_F, the loading rate R is

$$R = \frac{S}{\vartheta_F} \tag{5.26}$$

Then the current stress $\sigma(t)$ at moment t is $\sigma = Rt$, and integral (5.18) can be written for the ultimate damage fraction D_F in correspondence to expression (5.9) as

$$D_F = \tau_0^{-1} \int_0^{\vartheta_F} \exp\left(\frac{\alpha Rt - \varpi_0}{\widetilde{T}(t)}\right) dt \tag{5.27}$$

Pertinent to constant temperature, $\widetilde{T}(t) = \widetilde{T}$; that is

$$D_F = A^{-1} \int_0^{\vartheta_F} \exp\left(\frac{\alpha R}{\widetilde{T}} t\right) dt \tag{5.28}$$

where A has constant value

$$A = \tau_0 \exp\left(\frac{\varpi_0}{\widetilde{T}}\right) \tag{5.29}$$

Taking into account the ultimate condition $D_F = 1$, the solution of integral (5.28) can be presented as

$$\frac{\widetilde{T}}{A\alpha R}\left(\exp\left(\frac{\alpha S}{\widetilde{T}}\right)-1\right)=1 \qquad (5.30)$$

or

$$\frac{\widetilde{T}}{\tau_{S,T}}-\frac{1}{A}=\alpha R \qquad (5.31)$$

Note that for structural PMCs parameter A has an order of magnitude $A\sim10^{11}$ s. This means that for practical cases $A\gg\tau_{S,T}$. Therefore, the solution to (5.31) is approximately

$$\frac{\vartheta_F}{\tau_{S,T}}\widetilde{T}=\alpha S \qquad (5.32)$$

or, taking into account relation (5.12),

$$\alpha_0=\frac{\vartheta_F}{\tau_{S,T}}\widetilde{T} \qquad (5.33)$$

This relation essentially reveals the physical meaning of the dimensionless parameter α_0, a ratio of material endurance under constant-rate loading up to ultimate stress level S to that under the unchanging stress of ultimate level S. It is presumed that both these loading events are run under the same constant temperature $\widetilde{T}$. For standard temperature conditions, when $T=T_s$, relation (5.33) is

$$\alpha_0=\frac{\vartheta_F}{\tau_S} \qquad (5.34)$$

It should be emphasized that expressions (5.33) and (5.34) substantiate interrelation of material responses to different categories of loading conditions, reflected in the master curve notion (Miyano et al., 2005; Regel et al., 1972; Shkolnikov, 1995) introduced in Section 4.6.

Performing further transformations of (5.27), the relative value of ultimate strength beyond standard loading conditions $\widetilde{\sigma}_{u,T}$ can be expressed as a function of the length of constant-rate loading, as

$$\widetilde{\sigma}_{u,T}=\frac{\widetilde{T}}{\alpha_0-\widetilde{T}}\left(\alpha_0+\varpi_0\left(\frac{1}{\widetilde{T}}-1\right)-\ln\left(\widetilde{\vartheta}\right)-\ln\left(\widetilde{T}\right)-1\right) \qquad (5.35)$$

For standard temperature, as $\widetilde{T}=1$, this relation is

$$\widetilde{\sigma}_u=1-\frac{\ln\left(\widetilde{\vartheta}\right)}{\alpha_0-1} \qquad (5.36)$$

where $\widetilde{\vartheta}$ stands for the ratio of actual to standardized lengths of (test) loading.

Expressions (5.35) and (5.36) particularly embrace two major load cases: quasi-static short-term loading, usually employed for experimental determination of ultimate strength, and high-strain-rate-loading, as monotonous impulse loading.

To provide a well-grounded assessment of the influence of loading parameters to a material's ultimate strength, time-dependency parameters $\alpha_0 = 26.2$ and $\varpi_0 = 56.8$, experimentally determined by Lavrov and Shkolnikov (1991) for a marine-grade structural PMC, are used. These parameters are also used for a few other practical load cases presented below.

The analytical result is plotted in Figure 5.1, reflecting the influence of a loading length variation of $10^{-2} \leq \vartheta_F \leq 10^3$ s, which embraces both high-strain-rate and quasi-static short-term monotonic loadings.

As can be seen, variation of the loading rate makes a noticeable contribution to a measure of material load-bearing capability. In particular, loading length in the range of $1 \leq \vartheta_F \leq 10$ min (allowed by the standard ASTM D3039/D3039M-00 to produce specimen failure during tensile testing) would constitute a considerable portion of the test results dispersion, roughly corresponding to a variation coefficient of $CV \approx \pm 5\%$.

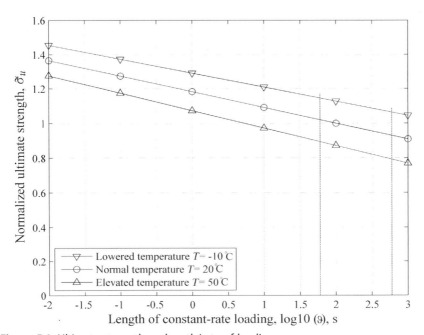

Figure 5.1 Ultimate strength vs. length/rate of loading.

The allowed variation of the test ambient temperature $T = T_s \pm 3$ °C further increases scattering of the test data, with extra $CV_T = \pm 2\%$. Jointly, the allowed variation of standardized testing conditions can produce scattering of test results consistent with the experimentally determined structural properties of PMCs.

> Just the allowed variation of the standard testing conditions can produce a scattering of test results consistent with customary experimentally determined structural properties of PMCs.

For this reason, the actual testing conditions should be carefully traced and taken into account to suppress the related dispersion of the test data. In particular, this can be done with utilization of the master curve notion as demonstrated in Section 4.6. Execution of this technique is particularly worthwhile for the testing of complex hull components, such as structural joints and/or large-/full-scale structural prototype sections, of which the load–bearing capability and hence the endurance under applied test loading are not completely predictable.

> The actual testing conditions should be carefully traced and taken into account to suppress the related dispersion of the test data. This is particularly important for testing of complex hull components and/or full-scale structural prototype sections, whose load-bearing capability and endurance under applied testing loading are not completely predictable.

5.3.2 Residual strength

Another useful application of kinetic-based analytical technique is characterization of residual strength. This physical testing is common for single large- or full-scale structure prototypes, the robustness of which needs to be verified for several load cases, such as short-term quasi-static loading and long-term fatigue loading. Accordingly, the structure undergoes two diverse test rounds, fatigue loading followed by after-fatigue short-term loading to failure, to define residual load-bearing capability of the tested structure, usually considered the threshold strength of that structure.

From the kinetic point of view, residual strength S_R is a function of ultimate strength expectation S, the preceding damage fraction D_0, accrued under introductory fatigue loading, and the length of after-fatigue static loading ϑ_R.

of $\tau_{Np} \leq \tau_{Nr} \leq 2\tau_{Np}$, where τ_{Np} is the endurance under pulse cycling with frequency and absolute values of peak stress are the same as those for reverse test loading.

The range itself is not significant enough to manifest a considerable extra value of fatigue strength for the tested material. Roughly, it is associated with only about ~2.5% of the ultimate strength for an event fully independent of load-bearing material constituents. For a practical case, the effect of reverse loading would be even lower because some involvement of the distinct material constituents in the shared load-bearing is practically unavoidable, and the relative peak stresses would never be equal at the different phases of a loading cycle.

Altogether, for all the given reasons, the influence of reverse cycling over pulse cyclic loading is insignificant for PMC serviceability characterization and may be neglected for design strength reconciliation of either the PMC or the hybrid structure.

> The influence of reverse cycling over pulse cyclic loading is insignificant for PMC serviceability characterization and may be neglected for design strength reconciliation of either PMC or hybrid structure.

Nevertheless, experimental examination of PMC performance under reverse cyclic loading should be implemented to verify validity of this logical deduction.

5.4 STRUCTURAL PERFORMANCE AT COMPLEX LOADING PROFILES

Extending the linear damage accumulation rule to a representation of distinct categories of loading exposure allows for characterization of a PMC's structural performance under a complex loading profile. In general, this can be expressed via the lifetime τ_{Σ} that sums up deteriorative influence of assorted operational exposures withstood by the PMC during service and/or test loading. Similar to that deduced by Boyzov (1997) for metal hull structures, it can be expressed as

$$\tau_{\Sigma} = \left(\sum_i^{n_i} \sum_j^{n_j} \sum_k^{n_k} \frac{P_{ijk}}{\tau_{ijk}} \right)^{-1} \tag{5.53}$$

Here τ_{ijk} denotes partial endurance of a PMC under ith category of operational exposure regarding jth load case at kth instance of ambient

conditions; n_i, n_j, and n_k are numbers of corresponding operational occasions; P_{ijk} is either the anticipated probability or designated portion of material exposure to be specified respectively for i, j, k operational events. Jointly, those portions of the material exposure make the sum

$$\sum_i^{n_i}\sum_j^{n_j}\sum_k^{n_k} P_{ijk} = 1 \tag{5.54}$$

The partial endurances τ_{ijk} are defined for each given load case employing integral (5.18) for the relevant ultimate damage fraction.

To take into account the 3D stress state typically experienced by a material within a ship structure under operational loading, failure criteria reflecting the 3D character of the stress state are employed with regard to all components $(\widetilde{\sigma}_{kl})_V, k,l = 1,2,3$.

The von Mises yield criterion represents the well-established standard for metal structures. Employing expression (4.2) introduced in Section 4.1, it is possible to determine a non-dimensional failure index Ψ_M for the metal part of a hybrid structure of interest.

The extended Norris–McKinnon criterion in its non-dimensional form (4.3) and (4.4) is a computing tool suitable for evaluation of the load-bearing capability of composite structural components undergoing a 3D stress state. The given version of the Norris–McKinnon criterion appears to be preferable to other available options for ultimate stress state analysis of a composite ship hull structure. As ascertained in Section 4.1, the extended Norris–McKinnon criterion provides a good match between analytical prediction and experimental data pertaining to the 3D stress state of PMCs with the relatively slender orthotropy (or transverse quasi–isotropy) typical for marine-grade PMCs. This criterion also supplies a consistent analytical result using just a few major strength characteristics of the utilized composite, alleviating the necessity for overly expanded test programs. Analogously to expression (4.2) for the von Mises criterion, the Norris–McKinnon criterion in forms (4.3) and (4.4) makes it possible to define a non-dimensional failure index Ψ_C relevant for a composite part of a hybrid structure. Jointly, the failure indexes Ψ_M, Ψ_C, determined for both primary parts of a hybrid structure, metal and composite, allow for identification and localization of critical areas within that hybrid structure with a unified failure index Ψ_{max} that is

$$\Psi_{max} = \max\{\Psi_M, \Psi_C\}_V \tag{5.55}$$

Both 3D criteria deal with linear relations of applied forces and induced stresses that correspond to the response of the vast majority of structures to operational loadings. This enables a favorable opportunity to represent the stress level in time-force relations, such as (5.9) and (5.18), directly via relative values of the applied loads $\widetilde{Q}$ instead of stress components $\{\widetilde{\sigma}\}$, as

$$\widetilde{Q} \equiv \frac{Q}{Q_u} = \{\widetilde{\sigma}\} \tag{5.56}$$

Due to this, characterization of serviceability of a structural PMC for known time-dependency parameters and given force-temperature exposure $\{Q(t), T(t)\}$ would result with ultimate load Q_u, corresponding to the given length of service $\tau_{Q,T}$, and vice versa.

Similarly to characterization of the ultimate parameters of load-bearing capability, allowed design loads can be determined, taking into account the required safety margin. In this case, the norm of design strength reconciliation for allowed stress components, expressed with relations (5.1) and (5.2), is transformed respectively to the allowed load of a given category of operational exposure Q_a,

$$\Psi_{max} \frac{Q_u}{f_s} \leq Q_a \tag{5.57}$$

Along with the opportunity to accurately characterize serviceability of composite and hybrid structural systems, the introduced analytical technique enables quantifiable prorating of destined loading exposures, enabling well-grounded optimization of a hybrid structure design consistent with a particular operational exposure.

> Along with the opportunity to accurately characterize the serviceability of composite and hybrid structural systems, the introduced analytical technique enables quantifiable prorating of destined loading exposures, enabling well-grounded optimization of a hybrid structure design for a particular operational exposure.

5.5 PRACTICAL APPLICATIONS

The introduced kinetic-based analytical technique is devised for and, in fact, is fairly instrumental in, serviceability characterization of composite/hybrid

ship hull structures undergoing essentially any operational loading. It is particularly effective for loading embodying force and ambient conditions, both altering in time.

For instance, a DSV's pressure hull during diving undergoes gradually intensified external pressure accompanied by decreased temperature of the surrounding seawater, particularly in tropical latitude oceans. A pipeline undergoing emergency overloading usually experiences concurrent surge of both internal pressure and temperature, which typifies another relevant operational occurrence.

Apparently, to properly evaluate serviceability of structures undergoing such load changing, parameters of the force and ambient conditions need to be taken into account. The following computational experiment illustrates this by quantitative assessment of the influence of loading profiles on the intensity of material deterioration of a hypothetical DSV pressure hull. The conceived loading schedule for the entire length of the hull's service consists of preliminary acceptance testing of the DSV hull in a hyperbaric tank, her sea trial, and provisional rounds of DSV service diving to assorted target ocean depths.

Note that external water pressure is only applied operational load to the pressure hull for clarity of the undertaken consideration. This external pressure is assumed to be linearly proportional to the ocean depth attained by the DSV in both her sea trial and destined service.

The loading rate for in-sea operations is supposed to be $R_s = 0.7$ m/s, to resemble customary submerging-surfacing velocity. For in-tank testing, for which loading rate is usually controlled by an employed pumping station, a realistic rate of $R_t = 50.0$ m/s is chosen.

The ambient temperature at the in-tank test loading is set at a constant $T = 20\ °C$. The changing ocean water temperature is presumed be within a range of $2.5 \leq T \leq 30\ °C$, typical for tropical latitudes. The temperature variation versus ocean depth is approximated with a power function that matches a temperature variation profile presented by Bergman (2011). Using the Celsius scale, this is

$$T = 27.5 \left(\frac{H}{1000} + 1 \right)^{-2.4} + 2.5 \tag{5.58}$$

where H is the depth of the ocean water in meters.

Table 5.2 specifies parameters of a loading schedule for a hypothetical composite pressure hull designed for a maximum depth of ocean operation

Table 5.2 Conceivable schedule for pressure hull loading

Load case	Target depth, m	Submerging-surfacing velocity, m/s	Stay on target depth, h	Ambient temperature	Diving rounds	Acquired damage fraction
Acceptance testing in hyperbaric tank						
1	7000	50.0	0	Room	2	0.006
Sea trial						
2	7000	0.7	0.1	Changeable	1	0.026
Operation						
3	≤2000	0.7	2	Changeable	5000	0.001
4	2001–3000	0.7	2	Changeable	1000	0.002
5	3001–4000	0.7	2	Changeable	500	0.018
6	4001–5000	0.7	2	Changeable	200	0.114
7	5001–5600	0.7	2	Changeable	100	0.302
8	5601–6000	0.7	2	Changeable	58	0.531
Total material deterioration during service life						**1.000**

of $H_{max} = 6000$ m and an ultimate depth of $H_U = 10,000$ m, which could cause hull failure.

Also presented are computed rates of the damage fraction of the structural material acquired under those discrete loading exposures as well as the total accumulated damage fraction.

Expressions (5.53) and (5.54) are employed to assess the hull's material deterioration. Concurrent change of force and temperature during repetitive submerging-surfacing operations is taken into account. Kinetic parameters $\alpha_0 = 26.2$; $\varpi_0 = 56.8$, specified in the reference Lavrov and Shkolnikov (1991) for a marine-grade PMC are again used. A safety factor of $f_s = 1$ is set for clarity of this analytical assessment of the influence of loading profile variability.

The graph in Figure 5.4 demonstrates the quantified deterioration (damage fraction) of DSV hull material prorated for each of the eight load cases.

Note that the water pressure applied for acceptance testing and for the sea trial prior to the pressure hull's entrance into service exceeds the maximum operation load, corresponding to $H_{test} = 7000$ m vs. $H_{max} = 6000$ m of ocean depth. This is to render common practice for acceptance testing of metal pressure hulls, and ensure sufficiency of hull robustness for the designated service.

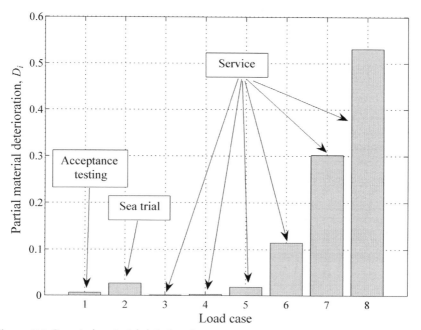

Figure 5.4 Prorated material deterioration.

Meanwhile, due to the enhanced sensitivity of structural PMCs to the parameters of loading, such customary overloading may notably affect the load-bearing capability and lifetime of a PMC. In the given particular case, the acceptance testing and sea trial together consume a roughly 3.5% portion of the initial load-bearing capability. This is disproportionately excessive, noticeably higher than for a marine-grade metal.

Overall, the performed assessment shows that the given analytical technique allows for notable advancement of the serviceability evaluation of a structural PMC without unduly complicating the conventional strength reconciliation routine for ship design. Moreover, this provides a well-justified ground for optimization of a composite/hybrid structural design with regard to the prorated impact of assigned loading or, on the other hand, allows for proper adjustment of the operational schedule for a particular structural design.

The introduced analytical technique allows for the notable advancement of serviceability evaluation of a structural PMC along with a well-justified ground for optimization of a composite/hybrid structure, without unduly complicating the conventional strength reconciliation routine of ship design.

The presented analytical approach is also instrumental for controlling the residual load-bearing capability of a real composite/hybrid hull structure in-service. In this instance, the initially assigned loading schedule needs to be substituted with actual experienced operational exposures.

5.6 EXPERIMENTAL VERIFICATION OF THE KINETIC-BASED APPROACH

A nonstandard target test program was executed to verify validity and effectiveness of the introduced kinetic-based analytical approach for serviceability characterization of structural PMCs. This program, partially reported in Lavrov and Shkolnikov (1991), comprised static and cyclic bending tests to failure of several sets of the 48- to 96-mm-thick coupons of a marine-grade epoxy parallel-diagonal (½, ¼, ¼) laminar GFRP.

Broadly ranging force-ambient conditions accompanied by meticulous control of actual parameters of the applied loading constituted the principal differentiating features of the performed tests from the conventional standards. Specifically, the tests comprised short-term monotonic loading with

stress rates in the range of $0.3 \leq R \leq 13.0$ MPa/s and long-term low–cyclic triangular-profile pulse loading with frequencies in the range of $0.03 \leq f \leq 0.5$ Hz. The peak stress of the cyclic tests was varied within a range of $0.6 \leq \tilde{\sigma}_{max} \leq 0.8$. The stress ratio $r = 0.1$ was applied for all cyclic tests. The ambient temperature during the short-term tests was maintained normal, whereas that of the cyclic tests was in the range of $19 \leq T \leq 30$ °C.

The factual test data were refined by applying the introduced analytical technique to unify the parameters of loading conditions, corresponding to a statutory testing baseline. The triangular pulse cyclic loading with frequency $f = 0.1$ Hz, stress ratio $r = 0.1$, and ambient temperature $T = 20$ °C constituted the principal parameters of the baseline selected. The actual peak stresses, i.e., those varied within a range of $0.6 \leq \tilde{\sigma}_{max} \leq 0.8$, were kept unchanged.

Two sets of the data acquired from the cyclic testing of the $1152 \times 96 \times 96$ mm material coupons are plotted in Figure 5.5 to illustrate fatigue performance of the tested PMC.

One set of data, denoted with triangular markers, embodies the original results actually gained from the tests. This set comprises both the short-term test results, treated as a half-cycle outcome of the cycling loading, and long-term test data reflecting the factual cyclic test results. The other data set,

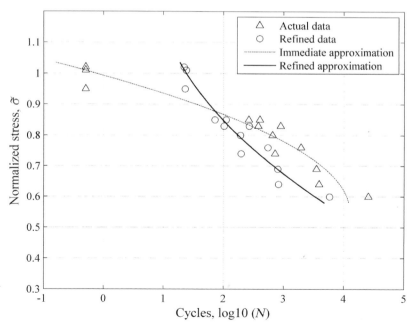

Figure 5.5 Actual and refined fatigue data, adapted from Shkolnikov (2007).

indicated with circular markers, embraces the original test data refined to match the parameters of the statutory test baseline.

Two approximation lines for both data sets are also plotted to represent performance trends relevant to these two sets of essentially the same experimental data.

The second order polynomial functions was used for both approximations, the appearance of which turned out to be quite different. The dotted line pertaining to the original test data is evidently bowed, whereas the solid line pertinent to the refined data is nearly straight. Although the tests were executed with a relatively minor variation of loading parameters, this did cause meaningful divergence of the original and refined data, perfectly illustrating the significance of accurate representation of the actual testing conditions for proper interpretation of experimentally derived data on PMC serviceability. Moreover, the refined data much better resemble the customary trend inherent to long-term performance of structural PMCs, associated with relations (4.27) and (5.15), than the original test data do. This fairly well testifies to the authenticity of the applied analytical technique and its validity for characterization of PMC serviceability for variable loading exposures. For this reason, utilization of refined data corresponding to unified testing parameters appears to be preferable to the routine relying directly on factual test data for characterization of PMC service performance.

> Utilization of refined data corresponding to unified testing parameters appears to be preferable for evaluation of a PMC's serviceability in lieu of routinely relying directly on the original test data.

As the presented analytical approach suffices for evaluation of a PMC's response to loading parameters differing from actual ones, it seems to be applicable for serviceability characterization of that PMC subjected to any given loading profile. To implement it, time-dependency parameters (τ_0; ϖ_0; α_0) of the material of interest as well as its ultimate strength at the relevant loading exposure all need to be known.

The above-deduced relations, including expressions (5.45) and (5.46), can be used to determine the required time-dependency parameters with regard to three (sets of) points on a fatigue diagram resulting from experimental examination of a PMC's performance. For instance, for fatigue cyclic pulse testing with a triangular loading profile, the relevant expression (5.45) is used for three test rounds, as

$$N_i = \frac{\tau_0 \alpha_0 \widetilde{\sigma}_i (1 - r_i) \exp\left(\dfrac{\varpi_0 - \alpha_0 \widetilde{\sigma}_i}{\widetilde{T}_i}\right)}{\vartheta_i \widetilde{T}_i \left(1 - \exp\left(\dfrac{\alpha_0 \widetilde{\sigma}_i}{\widetilde{T}_i}(r_i - 1)\right)\right)}, \quad i = 1,2,3 \qquad (5.59)$$

Note that execution of the related computing algorithm will need a few iterations due to slight non-linearity of Equation (5.59). This is a quickly convergent algorithm, and it typically takes not more than two iterations to gain a stable computational result.

To simplify the computing procedure, a few justifiable approximations may be applied. These include relation

$$\exp\left(\frac{\alpha_0 \widetilde{\sigma}_{max}}{\widetilde{T}}(r - 1)\right) \approx 0 \qquad (5.60)$$

which is acceptable for loading parameters of practical interest.

Because of this acceptability, Equation (5.59) comes into play

$$N_i = \frac{\tau_0 \alpha_0 \widetilde{\sigma}_i (1 - r_i) \exp\left(\dfrac{\varpi_0 - \alpha_0 \widetilde{\sigma}_i}{\widetilde{T}_i}\right)}{\vartheta_i \widetilde{T}_i} \qquad (5.61)$$

Moreover, in case $\widetilde{T}_{N1} = \widetilde{T}_{N2}$,

$$\alpha_0 = \frac{\widetilde{T}_{N1}}{\widetilde{\sigma}_{N2} - \widetilde{\sigma}_{N1}} \ln\left(\frac{N_1 \widetilde{\sigma}_{N2}(1 - r_2)\vartheta_{N1}}{N_2 \widetilde{\sigma}_{N1}(1 - r_1)\vartheta_{N2}}\right); \quad r_1 \neq 1 \qquad (5.62)$$

$$\varpi_0 = \frac{\alpha_0 \left(\widetilde{\sigma}_{N1}\widetilde{T}_{N3} - \widetilde{\sigma}_{N3}\widetilde{T}_{N1}\right) + \widetilde{T}_{N1}\widetilde{T}_{N3} \ln\left(\dfrac{N_1 \widetilde{\sigma}_{N3}(1 - r_3)\vartheta_{N1}\widetilde{T}_{N1}}{N_3 \widetilde{\sigma}_{N1}(1 - r_1)\vartheta_{N3}\widetilde{T}_{N3}}\right)}{\widetilde{T}_{N3} - \widetilde{T}_{N1}};$$

$$\widetilde{T}_{N3} \neq \widetilde{T}_{N1} \qquad (5.63)$$

$$\tau_0 = \frac{N_1 \vartheta_{N1} \widetilde{T}_{N1}}{\widetilde{\sigma}_{N1}(1 - r_1)} \exp\left(\frac{\alpha_0 \widetilde{\sigma}_{N1} - \varpi_0}{\widetilde{T}_{N1}}\right) \qquad (5.64)$$

To illustrate operability of the given algorithm, it is used to define the time-dependency parameters of Comeld-2 champion underwent in-plane tension test loading. The input comprises parameters of the applied loading and test results presented in Chapter 4. Specifically, this two-set input includes the following:

(1) Period of triangular pulse cycle $\vartheta_N = 7.5$ s; normal ambient tempera-
    ture $\widetilde{T}_N = 1.0$; and stress ratio $r = 0.1$.

(2) Length of monotonic short-term loading $\vartheta_3 = 100$ s; elevated ambient temperature $\widetilde{T}_u = 1.13$ (on the Kelvin scale); relative value of ultimate strength $\widetilde{\sigma}_3 = 0.92$.

Coefficients $B_N = 1.176$ and $\beta_N = 0.208$, defined in Section 4.8.3 to represent the approximate line plotted in Figure 4.15, are used to specify two experimental points of the $S - N$ fatigue diagram, $\{N_1 = 10; \widetilde{\sigma}_{N1} = 0.96\}$ and $\{N_2 = 10{,}000; \widetilde{\sigma}_{N2} = 0.35\}$.

The time-dependency parameters derived from this computation exercise are: $\alpha_0 = 13.0$; $\varpi_0 = 19.1$; $\tau_0 = 0.0087$ s. Similarly to the assessment presented in Chapter 5 for conceivable operation of a DSV composite pressure hull, the gained parameters are sufficient to characterize serviceability of the given Comeld-2 configuration at any anticipated operational (or test) loading situation.

Invariance of time-dependency parameters α_0; ϖ_0; τ_0 on direction of an applied force relative to orientation of an orthotropic PMC signifies another advantage of the presented approach over existing analytical techniques. To authenticate this notion, experimental data derived from studies specifically emphasizing "exfoliation" of the $S - N$ fatigue diagrams (Philippidis and Vassilopoulos, 1999; Sarkisian, 1984; Zakharov et al., 1967) have been explored.

In particular, absolute values of the fatigue strength of different PMC compositions for several characteristic load cases were transformed to the normalized view of dimensionless ratios of related fatigue and ultimate strength values. In the results, the original prominent "exfoliation" of the fatigue diagrams diminished for all the load cases explored, coming to a relatively narrow locus with scattering margins consistent with those normal for outcomes of fatigue testing of PMC coupons.

This result fairly well affirms the validity of the notion of invariance of time-dependency parameters of a PMC on orientation of the applied loading relative to the axes of the PMC orthotropy. This implies that the same fatigue diagram being obtained for one category of loading is sufficient to characterize fatigue performance of a given PMC for any load case, as ultimate strength relevant to the loading category of interest is also known.

The invariance of time-dependency parameters of a PMC on category and orientation of the applied loading relative to the axes of the PMC's orthotropy implies that the same fatigue diagram obtained for one load case is sufficient to characterize fatigue performance of a given PMC pertaining to any loading case, as ultimate strength relevant to a particular load case of interest is also known.

Owing to this encouraging introductory usage, the extensive test programs usually accompanying acquisition activities aimed at development of new naval platforms and allied certification of novel material systems can be truncated to just a few testing rounds. These will establish dependable experimental data sufficient for determination of time-dependency parameters and ultimate strength of the material of interest pertaining to all destined load cases.

For instance, such an amended test program might comprise:

- Short-term tests for all major load categories under normal ambient conditions to experimentally determine basic mechanical properties, including ultimate strength of that PMC along its axes of orthotropy
- Short-term tests under a primary load case and altered ambient conditions to evaluate the material's sensitivity to temperature and/or other environmental factors
- Fatigue tests under a primary load case at normal ambient conditions.

Certainly rigorous monitoring and tracing of actual testing parameters should accompany test execution. Any noticeable deviation of the testing parameters from their nominal statutory values needs to be documented to allow for appropriate refinement of the factual test data, similar to that reflected in Figure 5.5.

Particular attention should be paid to self-warming of test articles under cyclic testing. If that is unavoidable, either proper cooling-down or temperature gauging and documentation should be carried out.

On the whole, the presented experimental verification clearly demonstrates validity and suitability of the kinetic-based analytical approach for adequate interpretation of factual test data. This in turn enables accurate characterization of PMC serviceability within a composite/hybrid ship hull structure undergoing changing force-ambient loading exposures, either operational or testing, along with well-grounded specification of design allowables for the PMC.

> The presented experimental verification clearly demonstrates validity and suitability of the kinetic-based analytical approach for adequate interpretation of factual test data that enables accurate characterization of the serviceability of a PMC within a composite/hybrid ship hull structure undergoing changing force-ambient exposures, either operational or testing, along with well-grounded specification of design allowables for the PMC.

5.7 METHODOLOGICAL UPGRADE OF SERVICEABILITY EVALUATION

The improved accuracy of serviceability characterization and/or specification of design allowables attainable with an application of the introduced analytical technique enables meaningful advancement of design of heavy-duty naval structures. As with any analytical routine destined for structural analysis and strength reconciliation, the introduced technique is to fit and be congruent with conventional ship design practice. In fact, the presented analytical approach has a great deal of procedural similarity with the conventional analytical routine, without imposing any undue complications with regard to computations.

The technique is useful for traditional specification of design allowables followed by structural strength reconciliation, so for direct serviceability evaluation of that structure subjected to a loading exposure consistent with a given operational profile. This in turn allows for prorating of the influence of partial assigned loadings as well as prior-service acceptance testing expedient for effective structural optimization.

The input data on anticipated force-ambient conditions of operation may be presented in either probabilistic or deterministic terms, and the accuracy of the provided data will drive the accuracy of the pursued serviceability evaluation.

Essentially, the extended specification of anticipated loading parameters signifies the only added effort. As that is provided and as upgraded analytical evaluation of serviceability is implemented, that extra effort will be repaid with beneficial improvement of several key performance parameters of the designed structure. These include increased weight efficiency, enhanced reliability and safety of assigned operation, and amplified effectiveness of the construction cost.

Figure 5.6 delineates a flowchart of conventional design structural analysis and strength reconciliation with indication of the components to be upgraded.

In particular, the affected and/or added procedural steps include: specification of design load cases; evaluation of serviceability; prorating of design load cases; strength reconciliation; and structural design finalization.

As a heterogeneous hybrid structure embodies at least two diverse material systems, metal and composite, they must satisfy requirements of the design strength reconciliation. The existing guidelines for warship design

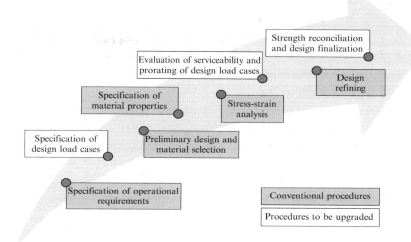

Figure 5.6 Projected upgrade of design strength reconciliation.

impose these requirements and describe the reconciliation procedure separately for each material category, metal and composite.

As discussed above, this approach is deficient for evaluation of interaction of diverse material systems within a hybrid structure that might be significant in some instances. These include but are not limited to different thermal expansion of the dissimilar material systems and distinct fatigue performance of those material systems under changing parameters of a force-ambient operational exposure. The presented analytical technique is capable of addressing the peculiarities of the structural behavior of hybrids, taking into account the envisaged interactions of the dissimilar materials with proper accuracy.

Regarding the opportunity to considerably truncate experimental programs that normally accompany acquisition activities, application of the introduced analytical technique allows for sizeable reduction of the range, length, and cost of those testing programs.

Also, the technique is worthwhile for analytical empowerment of the health monitoring of composite and hybrid structures in-service enabling a proper representation of actually undergone operational exposures. In this case, the critical areas of an exposed structural component can be accurately ascertained.

Overall, the presented kinetic-based technique seems to be an effective and dependable analytical tool suitable for adequate interpretation of test

results; accurate serviceability characterization; well-grounded design optimization; and in-service health monitoring pertaining to assorted heavy-duty structural systems utilizing PMC components.

The presented kinetic-based technique seems to be an effective and dependable analytical tool suitable for adequate interpretation of test results; accurate serviceability characterization; well-grounded design optimization; and in-service health monitoring pertaining to assorted heavy-duty structural systems utilizing PMC components.

Several computer codes for the MatLab environment provided in Appendix illustrate operability of the presented analytical technique.

REFERENCES

ABS, 2007. 109: hull structures and arrangements, Section 3, Plating. In: Hull Construction and Equipment, Chapter 2, Guide for Building and Classing High Speed Naval Craft, Part 3. ABS, Houston, TX, 243 pp.

ASTM D3039/D3039M-00, 2000. Standard Test Method for Tensile Properties of Polymer Matrix Composite Materials.

Bergman, J., 2011. Temperature of ocean water. Windows to Universe, February 16. Available from www.windows2universe.org/earth/Water/temp.html.

Boyzov, G.V., 1997. A concept of hull structure fatigue strength reconciliation. In: The Second International Conference and Exhibition on Marine Intellectual Technologies, Sect. 4, Intellectual Technologies in Applied Research, Russia, SPb, pp. 274–278.

D 3479/D 3479M-96, 2002. Standard Test Method for Tension-Tension Fatigue of Polymer Matrix Composite.

DNV, 2010. Offshore Standard, Composite Components, DNV-OS-C501. Det Norske Veritas, October, 164 pp.

DNV, 2012. Rules for Classification of High Speed, Light Craft and Naval Surface Craft. Pt. 3 4(1), Det Norske Veritas, July, 49 pp. Available from http://exchange.dnv.com/publishing/RulesHSLC/2012-07/hs304.pdf.

Echtermeyer, A.T., Ronold, K.O., Breuil, O., Palm, S., Hayman, D., Noury, P., Osnes, H., 2002. Project offshore standard composite components. Technical Report No. 2002-0124, Rev.1, Det Norske Veritas, February 8, 409 pp. Available from http://www.bsee.gov/Research-and-Training/Technology-Assessment-and-Research/tarprojects/300-399/343AA.aspx.

Greene, E., 1997. Design guide for marine applications of composites. Final Report, SSC-403, 337 pp. Available from http://www.shipstructure.org/pdf/403.pdf.

Lavrov, A.V., Shkolnikov, V.M., 1991. Experimental research of fatigue of thick-walled FRP structures, structural application of PMC. Works of the Krylov Scientific-Technical Society, 510, SPb, Russia, pp. 4–15 (Экспериментальное Исследование Усталости Стеклопластика Больших Толщин при Циклическом Изгибе).

Michaelov, G., Kharrazi, M., Momenkhani, K., Sarkani, S., Kihl., D.P., 1997. Fatigue behavior and damage accumulation of FRP laminates and joints. Technical Report, NSWCCD-TR-65-97/36, NSWCCD, 159 pp. Available from www.dtic.mil/cgi-bin/GetTRDoc?AD=ADA342353.

Miner, M.A., 1945. Cumulative damage in fatigue. Journal of Applied Mechanics 12, A156–A164.

Miyano, Y., Nakada, M., Sekine, N., 2005. Accelerated testing for long-term durability of FRP laminates for marine use. Journal of Composite Materials 39 (1), 5–20.

Palmgren, A.Z., 1924. Durability of ball-bearings. Zeitschrift des Vereines Deutscher Ingenieure 68, 339–345 (Die Lebensdauer von Kugellagern).

Philippidis, T.P., Vassilopoulos, A.P., 1999. Fatigue strength prediction under multiaxial stress. Journal of Composite Materials 33 (17), 1578–1598.

RD 5.1186-90, n.d. Composite Hulls and Hull Structures. Design Rules and Methods of Strength Analysis (Корпуса и Корпусные Конструкции из Стеклопластика. Правила Проектирования и Методические Указания по Расчетам Прочности), USSR.

Regel, V.R., Slutsker, A.I., Tomashevsky, E.E., 1972. The kinetic nature of the strength of solids. Progress of Physical Sciences 106 (2), 193–228, Moscow, Russia (Кинетическая Природа Прочности Твёрдых Тел).

Sarkisian, N.E., 1984. Anisotropy of fatigue strength of PMCs. Mechanics of Structures of PMCs, Novosibirsk, Science – Siberian Branch, USSR, pp. 92–97 (Анизотропия Усталостной Прочности Композитных Материалов).

Shkolnikov, V.M., 1995. Principles of computational life prediction of composite structures for practical application. In: Questions of Material Science, 3, Russia, SPb, CSRI CM 'Prometey', pp. 30–38 (Принципы Расчётного Прогнозирования Прочности Корпусных Конструкций из Полимерных Композитов в Аспекте Практического Приложения).

Shkolnikov, V.M., 2007. Serviceability characterization of composite storage tank, PVP2007-26099. In: Proceedings of ASME Pressure Vessels & Piping Conference and Eighth International Conference on CREEP at Elevated Temperature (PVP2007/CREEP8), San Antonio, Texas, pp. 483–492.

Smirnova, M.K., Paliy, O.M., Spiro, V.E., 1984. Principles of strength norms specification for composite structures. Mechanics of Polymer Materials, 882–887, March, Riga, Latvia (О Принципах Назначения Норм Прочности для Конструкций из Композитных Материалов).

Zakharov, V.A., Rivkind, V.N., Ragulina, T.L., Sidorenkova, A.N., 1967. Strength and deformation characteristics of GFRP under long-term static and cyclic tension, properties of polyester GFRPs and methods of their control. Works of Krylov Scientific-Technical Society, Leningrad: Shipbuilding, USSR, pp. 125–166 (Прочностные и Деформационные Характеристики Стеклопластиков при Длительном Статическом и Циклическом Растяжении).

Zhurkov, S.N., 1984. Kinetic concept of the strength of solids. International Journal of Fracture December 26 (4), 295–307.

Zhurkov, S.N., Tomashevsky, E.E., 1966. Time dependence of strength under various loading regimes. In: Physical Basis of Yield and Fracture, Conf. Proc., Institute of Physics, London, 1966, p. 200.

CHAPTER 6

Prospective Investigations

A few more engineering aspects of prospective utilization of the most promising Comeld-2-based material-transition technology deserve further investigation in the effort to expand beneficial utilization of Comeld-2 for assorted hybrid structural systems. The developmental targets include provision of electromechanical disconnection of a conductive PMC and metal laps within the material-transition structure to prevent galvanic corrosion of the metal part at the composite-metal interface; exploration of a laser option for Surfi-Sculpt protrusion, enabling further cost effectiveness of Comeld-2's industrial application; and prospective expansion of Comeld-2 technology beyond its naval implication, in particular for the wind-power generation industry and assemblage of the large modular blades of wind turbines. Prospective realization of all outlined innovative technologies seems feasible and worthy of elaboration. The following elucidates these investigative targets in depth.

6.1 PREVENTION OF GALVANIC CORROSION

Fiber-reinforced structural PMCs are commonly corrosion resistant. Most are practically nonconductive and can be joined to metals without fear of galvanic corrosion. The problem arises when an electrically conductive PMC, such as CFRP, is applied to a composite section of a hybrid hull at the area of its transition to the metal.

In general, a rationally designed hybrid hull may have both types of composite materials, conductive and nonconductive, structurally integrated with the primarily metal hull. The above-introduced Comeld-2 technology, although structurally superior to other hybrid joining options, does not prevent the galvanic corrosion that may develop when a conductive PMC is used in conjunction with the metal hull, analogous to that inherent to both bolted and bonded-bolted joints.

To address this issue, the original Comeld-2 configuration presented in Chapters 3 and 4 should be modified by incorporating a nonconductive material in the interface of metal and electro-conductive composite parts,

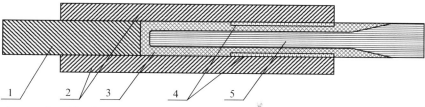

1 – metal middle plate
2 – metal lap plates
3 – non-electro-conductive PMC element
4 – partly protruded metal contact surface
5 – electro-conductive element

Figure 6.1 Modified Comeld-2 structure.

thereby preventing electrical transmission that would otherwise induce galvanic corrosion in the marine environment.

The technical solution presented in the Shkolnikov patent (2011) provides a proper modification of the original Comeld-2 configuration. The diagram presented in Figure 6.1 adapted from that patent outlines the suggested Comeld-2 modification.

In this case, a material-transition structural component is added with an intermediate non–electrically conductive element, to prevent the direct electrical contact associated with possible corrosion between the metal and the base PMC part that is presumably electrically conductive. That intermediate insert could be, for example, GFRP, or a rubber or ceramic matrix composite.

A rational combination of two Comeld-2 configurations, the ordinary type and the modified type outlined in Figure 6.1, might be preferable, providing enhanced flexibility in the multimaterial hull layout and thereby maximizing benefits of the composite application, meeting specific design and operational requirements.

While functionally beneficial, the use of two or more PMC material systems along with two distinct Comeld-2 configurations may cause manufacturing complications associated with intensified labor and extra expense. Presumably, this would be recompensed with the anticipated improvement of key performance parameters of the hybrid hull, pertaining to structural weight savings, higher payload, and fuel economy, among others.

To alleviate the concern about affordability and cost effectiveness, a cost assessment and tradeoff study on the multimaterial hybrid hull structure should be executed with regard to the respective implications.

To verify the feasibility and structural efficiency of the outlined modification for multimaterial Comeld-2 technology, relevant computer simulations as well as structural and galvanic resistance testing should be carried out.

6.2 LASER-BASED SURFI-SCULPT

While the Comeld-2 technology is considerably superior in structural efficiency and cost effectiveness to all existing heavy-duty hybrid joining options, the EB Surfi-Sculpt (Buxton & Dance, 2005 and Dance & Kellar, 2004) being employed is a relatively expensive operation that might inhibit a broad industrial application of Comeld-2. In addition, the necessity to operate with the EB in a vacuum limits the geometric complexity and size of work specimens that can be processed.

An alternative inexpensive laser-based protrusion technique, not requiring a vacuum operating environment, appears capable of producing a Surfi-Sculpt pattern similar to that selected for the Comeld-2 champion option. The principal advantage of the laser Surfi-Sculpt over its EB counterpart is the opportunity to perform protrusion at atmospheric pressure, either with or without a shielding gas, depending upon the material being processed.

Comparatively recently, the laser-based protrusion process has been demonstrated by Hilton & Nguyen (2008) and Blackburn & Hilton (2010). Macroscopic surface features were produced in several metallic materials, using high-powered multimode solid-state lasers in combination with a beam scanner. The preliminary trials have resulted in the initial establishment of key process parameters for protrusion of the simplest features, followed by the integration of these features into more geometrically complex features, and finally arrays of these features.

Blackburn & Hilton (2010) demonstrate features produced as a result of a laser Surfi-Sculpt trial on a 316 stainless steel bar, the pattern of which is somewhat like that of the champion Comeld-2 protrusion. Despite a certain shape similarity, the new laser-based pattern is yet beyond the required parameters for hybrid joint application.

In addition, as ascertained by Blackburn & Hilton (2010), along with the promising results of the preliminary trials, they have also shown that the thermal behavior of the work piece needs to be properly managed to ensure that nominally identical features are produced. Features within the array need be built simultaneously in order to optimize the build-rate of the features. This is not possible with existing beam scanning unit software designed for laser marking and/or remote processing applications.

For these reasons, the laser-based Surfi-Sculpt technique needs further elaboration in relation to use with hybrid joints.

The conceived technical approach is based on adaptation of a low-power laser beam to a metal treatment technique capable of producing surface features suitable for a robust bonded-pinned hybrid joint. This would allow for the practically infinite length of the metal protrusion to be consolidated with a composite component of a sizable material-transition structure without any of the extra welding operations needed for the original EB-based Comeld-2 implementation, thereby increasing performance and reducing the cost of the hybrid structural system.

It should be noted that the envisaged technological advancement is associated with a certain technical risk related to the capability of protruding the metal substrate with a laser beam to get a pattern suitable and sufficient for structural hybrid joint application. While the aimed-for metal protrusion is potentially feasible, it could be tricky to gain a desired pattern. To mitigate the risk, several different treatment types based on a down-selected protrusion pattern of a given steel grade should be examined, providing the opportunity to pick the right protrusion option.

Corresponding to these premises, the principal objectives of the initial stage of the projected investigation on laser Surfi-Sculpt should consist of:

- Down-selection of the protrusion patterns similar to those chosen for the original EB-based champion Comeld-2 and suitable for the laser-based processing
- Devising of a laser-based protrusion technique for the marine-grade steel (HSLA-65 family) alloy and/or other metal alloys of potential interest
- Running of fabrication trials, producing protrusion samples corresponding to the down-selected protrusion patterns
- Techno-economic assessment of the laser-based protrusion technique being devised.

As the projected investigation delivers a positive outcome sufficient for heavy-duty hybrid joining, it would further benefit utilization of the hybrid hull concept for naval vessels, applying the laser-based Comeld-2 material-transition structures.

It is envisioned that for end users, the cost of the laser solution would be an order of magnitude less than that of the EB system. Ultimately, the targeted technology would substantially increase the market value of the hybrid joining technology and hybrid structure concept implementation, on the whole.

6.3 COMELD-2 NON-NAVAL APPLICATIONS

Although Comeld-2 joining technology is primarily destined for the Navy's hybrid hull concept, it seems suitable and potentially beneficial for assorted nonmilitary applications. Mainly, those would involve heavy-duty applications for high-speed crafts, airplanes, automobiles, structural components of offshore floating platforms, pipelines for natural gas distribution and transmission, modular blades and towers of mega-sized wind turbines, and so on.

The potential benefits for sea/air/ground vehicles and floating platforms are largely similar to those for warship application, whereas the benefits associated with wind turbines include rational realization of a modular blade concept favorable for many instances of wind turbine manufacturing, erection, and operation.

In particular, the energy efficiency of a wind turbine grows significantly as its sweep area increases. This is due to the amount of the power converted from wind into rotational energy of the turbine P, which is proportional to the product of the square of the blade length L and the cube of the wind velocity v, i.e.,

$$P \propto L^2 v^3 \tag{6.1}$$

In reality, the positive effect of turbine enlargement is even greater because it is allied with increased height of the tower, implying access to faster winds.

Due to these factors, the trend is to maximize turbine size as much as technically possible. However, simply lengthening a blade without changing the fundamental design seriously complicates manufacturability, transportability, and on-site assembly of the turbine. The modular blade design is a way of enabling the benefits associated with turbine blade and/or tower enlargement while lessening the negative impact (Marsh, 2014).

The straightforward solution of utilizing plain adhesive bonding is limited, though, because of the intensity of the applied loading, the enormity of the bonding area, and hence the excessiveness of the molding operations necessary to assemble blades of such a mega turbine on the site of its installation.

A more practical approach is associated with mechanically bolted modular sections of either the blade or tower, as is demonstrated by Anon (2005) and Marsh (2014).

However, this option is also not ideal as it is associated with labor-intensive hole drilling and bolting operations at the site of mega turbine

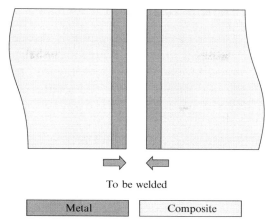

To be welded

| Metal | Composite |

Figure 6.2 Modular blade – shipbuilding approach.

erection, considerably affecting both the cost and quality of the turbine assembly.

An application of the Comeld-2 technology effective for a heavy-duty ship hull hybrid structure seems to be a suitable alternative for mega wind turbines. In this case, the conventional ship hull assembly approach based on welding of large hull sections would need to be executed for composite modules of either the turbine's blades or its tower. Outfitting with a metal skirt extended from the Comeld-2 structure could be done, as delineated in Figure 6.2.

Conceivably, this will allow for reduction of the complexity of the assembly intrinsic to fabrication and erection of modular wind mega turbines. Apparently, certain feasibility and tradeoff studies should accompany fulfillment of this projected development.

REFERENCES

Anon, 2005. 30m split rotor blade, MEGAWIND. Available from, http://www.cres.gr/megawind/split_rotor_blade.htm.

Blackburn, J., Hilton, P., 2010. The Generation of Autogeneous Surface Features Using a Low Power Laser Beam. TWI, Granta Park, Great Abington, Cambridge, 8 pp., Paper 1704.

Buxton, A.L., Dance, B.G.I., 2005. Surfi-Sculpt^{TM} – Revolutionary surface processing with an electron beam. In: Proceedings of ASM International ISEC Congress, St Paul, MS, 1–3 August.

Dance, B.G.I., Kellar, C., 2004. Workpiece Structure Modification. Number WO 2004/028731 A1, International Patent Publication. Available from, https://data.epo.org/publication-server/rest/v1.0/publication-dates/20101103/patents/EP1551590NWB1/document.pdf.

Hilton, P., Nguyen, L., 2008. A new method of laser beam induced surface modification using the Surfi-Sculpt® process. In: Proceedings of the 3rd Pacific Conference on Applications of Lasers and Optics, April, Beijing, China, 6 pp.

Marsh, G., 2014. Greater role for composites in wind industry. Reinforced Plastics 20–24, January/February.

Shkolnikov, V.M., 2011. Structural Component for Producing Ship Hulls, Ship Hulls Containing the Same, and Method of Manufacturing the Same. Patent 8,020,504 B2, USA.

APPENDIX: MATLAB CODES ON SERVICEABILITY CHARACTERIZATION

7.1 ULTIMATE STRENGTH VERSUS LENGTH OF LOADING AND TEMPERATURE

```
%Ultimate strength vs loading duration to failure &
%temperature
clear
clf
%Kinetic parameters
ao=26.2;
wo=56.8;
%Standard length of loading, s
ts=100;
%Standard temperature, K
Ts=273.15+20;
for j=1:3
%Temperature variation
    T=Ts-60+30*j;
    T=T/Ts;
    %Variation of length of loading, s
    for i=1:6
    t=10^(i-3);
    su=T/(ao-T)*(ao+wo*(T^-1-1)-log(t/ts)-log(T)-1);
    Su(i)=su;
    tl(i)=log10(t);
    end
     if T<1
     plot(tl,Su,'b-.')
     hold on
     end
     if T==1
     plot(tl,Su,'r-')
     hold on
```

```
      end
      if T>1
      plot(tl,Su,'g:')
      end
   end
xlabel('Length of constant-rate loading,log10(t),sec')
ylabel('Ultimate strength relative to standard')
legend('Lowered temperature T=-10C','Normal temperature
...T=20C','Elevated temperature T=50C',3)

grid
```

7.2 RESIDUAL STRENGTH

```
%Residual strength vs. length of monotonic loading and
%temperature
clear
clf
%Kinetic parameters
ao=26.2;
wo=56.8;
%Standard length of loading, s
ts=100;
%Actual relative length of loading, s
tr=100;
tr=tr/ts;
%Standard temperature, K
Ts=273.15+20;
for j=1:3
%Temperature alteration
    T=Ts-20+10*j;
    T=T/Ts;
    for i=1:1001
    do=.001*(i-1);
    sr1=T/(ao-T)*(ao+wo*(T^-1-1)-log(tr)-log(T)-1);
    sr=sr1+log(.9999-do)/ao;
    Do(i)=do;
    Sr(i)=sr;
    end
```

```
      if T<1
      plot(Do,Sr,'b-.')
      hold on
      end
      if T==1
      plot(Do,Sr,'r-')
      hold on
      end
      if T>1
      plot(Do,Sr,'g:')
      hold on
      end
end
xlabel('Preceding damage fraction, Do')
ylabel('Ratio of residual to ultimate strength, Rs')
legend('Lowered temperature T=10C','Normal temperature
...T=20C','Elevated temperature T=30C',3)
grid

axis([ 0 1 0 1.2])
```

7.3 CYCLIC LOADING

```
%Comparison of triangular & haversine cyclic &
%protracted loadings
clear
clf
%Kinetic parameters
wo=56.8;
ao=26.2;
to=1e-13;
%Cycle period, sec
te=6;
r=0;
for i=1:12
    %Stress level
    sm=.1*i;
    Sm(i)=sm;
    ps=ao*sm;
    %Protracted
```

```
      td=to*exp(wo-ps);
      ltp(i)=log10(td);
      %Triangular
      tt=td*ps*(1-r)/(1-exp(ps*(r-1)));
      ltt(i)=log10(tt);
      %Haversine
      ts=to*exp(wo-ps*(1+r)/2)/besseli(0,ps*(1-r)/2);
      lts(i)=log10(ts);
end
plot(ltt,Sm,'b:',lts,Sm,'g-.',ltp,Sm,'m-')
xlabel('Durability under cyclic/protracted loading,
...log10(), s')
ylabel('Relative stress')
legend('Triangular cycle','Haversine cycle',
...'Protracted loading')
grid

axis([ 0 9 0 1.2])
```

7.4 DSV DIVING

```
%Accumulated damage fraction
clear
clf
%Kinetic parameters
ao=26.2; wo=56.8; to=1e-13;
%Ultimate ocean depth, m;
S=10000;
%Target depths, m
Ht=[ 7000 7000 2000 3000 4000 5000 5600 6000];
%Loading rate/velocity, m/s
v=[ 50 .7 .7 .7 .7 .7 .7 .7];
%Divings
nd=[ 2 1 5000 1000 500 200 100 58];
%Stay under sustained loading, s
tS=[ 0 .1 2 2 2 2 2 2]*3600;
%****** Preliminary lab test ******
%The testing is presumed to correspond to a triangular
%cycle in a hyperbaric tank with room temperature
%TL=20C and rate
```

```
%of loading-unloading 50m/s for relative max pressure
%pm=0.7, e.g., target depth Ht=7000m
%Cycle period, sec and pressure range
te=2*Ht(1)/v(1);
r=0;
ps=ao*Ht(1)/S;
%Temperature, C & K
TL=20;
TL=(273.15+TL)/293.15;
%Deterioration under test triangle cycles
Dc(1)=nd(1)/((to*ps*(1-r)*exp((wo-ps)/TL)/TL
.../(1-exp(ps*(r-1)/TL)))/te)+tS(1)*nd(1)/to/
...exp((wo-ps)/TL)
%****** Sea trail & operation ******
%Integration
dc=0;
DS=0;
m=100;
for j=1:7
    j1=j+1;
    % Target depth, m
    ht=Ht(j1);
    % Length of submerging/surfacing, s
    ts=Ht(j1)/v(j1);
    % Time step, m
    dt=ts/m;
    ds=0;
  for i=1:m+1
    %Current depth, m
     Hc=v(j1)*dt*(i-1);
    %Current pressure
    p=Hc/S;
    ps=ao*p;
    %Current temperature, C
     Tc(i)=27.5*(Hc/1000+1)^-2.4+2.5;
    %Current relative temperature, K
     Tk=(Tc(i)+273.15)/293.15;
    %Integration for submerging & surfacing
     ds=ds+2*dt/to/exp((wo-ps)/Tk);
```

```
    %Integration for staying on target depth
    if i==m+1
      ds=ds+tS(j1)/to/exp((wo-ps)/Tk);
    end
%     Control of temperature distribution at ocean depth
%     plot(Tc,HC)
%     hold on
%     xlabel('Ambient temperature, C')
%     ylabel('Ocean depth, m')
%     grid
%     axis ij
    end
Dc(j1)=ds*nd(j1);
end
for j=1:8
DS=DS+Dc(j);
end
Dc=Dc
DS=DS
bar(Dc)
xlabel('Load case')
ylabel('Partial material deterioration')
grid

axis ([0 9 0 .6])
```

GLOSSARY/ABBREVIATIONS

Adherend a material or part that is held to another by an adhesive

AEM/S advanced enclosed mast/sensor system

ATD advanced technology demonstration

Comeld bonded-pinned composite-to-metal joining technology

Comeld-2 Comeld joint configuration with metal double lap

CRFP carbon-fiber-reinforced plastic

CTC Concurrent Technologies Corporation, Johnstown, Pennsylvania

CTD cool temperature dry

CTE coefficient of thermal expansion

DCNS naval defense company based in France, one of Europe's leading shipbuilders

DDG NATO standard designation for guided missile destroyers. Many of these vessels are also equipped to carry out antisubmarine, anti-air, and antisurface operations.

DNV Det Norske Veritas, a Norwegian classification society

DSV deep-submergence vehicle

EB electron beam

ESS engineered syntactic systems

ETD elevated temperature dry

ETW elevated temperature wet

EWI Edison Welding Institute, Columbus, Ohio

FE finite element

FEA finite element analysis

FMV Swedish Defence Materiel Administration

FRP fiber-reinforced plastic

FSU Former Soviet Union

GRFP glass-fiber-reinforced plastic

HII Huntington Ingalls Industries

KSRC Krylov State Research Centre, St. Petersburg, Russia

KVASI *Kockums* vacuum-assisted sandwich infusion

LPD An amphibious transport dock, also called a landing platform/dock, which is an amphibious warfare ship that embarks, transports, and lands elements of a landing force for expeditionary warfare missions.

MANTEC Navy Manufacturing Technology Program

MCMV mine countermeasures vessel

NSWCCD Naval Surface Warfare Center, Carderock Division, West Bethesda, Maryland

ONR Office of Naval Research, Arlington, Virginia

PMC polymer matrix composite

PMI polymethacylimide

Prepreg a reinforcing fibrous material pre-impregnated with a resin system

PVC polyvinyl chloride

QA quality assurance

RCS radar cross section
ROM rough order-of-magnitude
RTD room temperature dry
RTW room temperature wet
SCRIMP Seemann composites resin infusion molding process
Surfi-Sculpt metal protrusion process employing power beam
TDP technology demonstration grillage panel
TRL technology readiness level
TSI Triton Systems, Inc., Chelmsford, Massachusetts
TWI The Welding Institute, Cambridge, United Kingdom
USNA US Naval Academy, Annapolis, Maryland
UV ultraviolet
VIP closed-mold vacuum-assisted infusion material processing
WMT&R Westmoreland Mechanical Testing & Research, Inc., Youngstown, Pennsylvania

INDEX

Note: Page numbers followed by *f* indicate figures and *t* indicate tables.

Experimental determination of the residual strength of a composite/hybrid structure may be as informative as that for routine testing of a metal structure only in the event that all significant testing parameters are properly counted. Otherwise, residual strength testing may produce a quite misleading result.

5.3.3 Cyclic fatigue loading

Fatigue performance under cyclic loading represents one more practical case that can be analyzed in detail utilizing a closed-form solution of integral (5.18) and the kinetic-based analytical concept. In general, this can be expressed relative to the number of loading cycles, N_F, endured by a structural material up to its failure, as

$$D_F = N_F \int\limits_0^\vartheta \frac{dt}{\tau(\sigma(t), T(t))} \tag{5.42}$$

For two conventional waveforms of pulse loading, triangular (broken-line) and haversine, integral (5.42) has closed-form solutions when the ambient temperature remains unchanged, i.e., $\widetilde{T}(t) = \widetilde{T}$.

The loading cycles of these two waveforms can be described as follows.

• For a triangular pulse cycle, the induced stress within a cycle is

$$\widetilde{\sigma} = \begin{cases} \widetilde{\sigma}_{\max}\left(r + (1-r)\dfrac{2t}{\vartheta}\right), 0 \leq t \leq 0.5\vartheta \\ \widetilde{\sigma}_{\max}\left(1 - (1-r)\dfrac{2t-\vartheta}{\vartheta}\right), 0.5\vartheta < t \leq \vartheta \end{cases} \tag{5.43}$$

• For a haversine pulse cycle, it is

$$\widetilde{\sigma} = 0.5\widetilde{\sigma}_{\max}\left(1 + r + (1-r)\sin\,2\pi\left(\frac{t}{\vartheta} - 0.25\right)\right), 0 \leq t \leq \vartheta \tag{5.44}$$

Here $\widetilde{\sigma}_{\max}$ is normalized peak stress; ϑ is the period of the cycle; and $r = \frac{\sigma_{\min}}{\sigma_{\max}}$ is the stress ratio of the cyclic loading.

Corresponding to the given loading conditions, the ultimate number of cycles of the triangular pulse loading is

$$N_t = \frac{\tau_0 \alpha_0 \widetilde{\sigma}_{\max}(1-r)\exp\left(\dfrac{\varpi_0 - \alpha_0 \widetilde{\sigma}_{\max}}{\widetilde{T}}\right)}{\vartheta \widetilde{T}\left(1 - \exp\left(\dfrac{\alpha_0 \widetilde{\sigma}_{\max}}{\widetilde{T}}(r-1)\right)\right)} \tag{5.45}$$

That of the haversine loading is

$$N_s = \frac{\tau_0 \exp\left(\dfrac{2\varpi_0 - \alpha_0 \widetilde{\sigma}_{max}(1+r)}{2\widetilde{T}}\right)}{\vartheta I_0\left(\dfrac{\alpha_0 \widetilde{\sigma}_{max}(1-r)}{2\widetilde{T}}\right)} \tag{5.46}$$

where $I_0(\ )$ is the modified Bessel function of 0 order.

The given expressions are similar to the conventional view of the empirical expression (4.27) and can be converted into that. Parameters β_N, and B_N of relation (4.27), along with customary characterization of material fatigue performance under given loading conditions, would also embody parameters of the applied cyclic loading.

For a triangular cycle, those parameters would be

$$\beta_N = \frac{1}{M\left(\dfrac{\alpha_0}{\widetilde{T}_N} - 1\right)} \tag{5.47}$$

$$B_N = \beta_N \left(\log_{10}\left(\frac{\tau_0 \alpha_0 (1-r)}{\vartheta \widetilde{T}_N}\right) + M\left(\frac{\varpi_0}{\widetilde{T}_N} - 1\right)\right) \tag{5.48}$$

where $M = \frac{1}{\ln(10)}$.

Note that two justifiable approximations are used here to avoid a numerical representation of the given solution. These are

$$\ln(\widetilde{\sigma}_{max}) \approx \widetilde{\sigma}_{max} - 1, \text{ for } \sim 0.4 \leq \widetilde{\sigma}_{max} \leq 1.3 \tag{5.49}$$

and

$$\exp\left(\frac{\alpha_0 \widetilde{\sigma}_{max}}{\widetilde{T}}(r-1)\right) \approx 0 \tag{5.50}$$

To characterize the lifetime of a PMC undergoing long-term cyclic loading, expressions (5.45) and (5.46) are presented as follows.

- For the triangular pulse cycle

$$\tau_{Nt} \equiv N_t \vartheta = \frac{\tau_0 \alpha_0 \widetilde{\sigma}_{max}(1-r)\exp\left(\dfrac{\varpi_0 - \alpha_0 \widetilde{\sigma}_{max}}{\widetilde{T}}\right)}{\widetilde{T}\left(1 - \exp\left(\dfrac{\alpha_0 \widetilde{\sigma}_{max}}{\widetilde{T}}(r-1)\right)\right)} \tag{5.51}$$

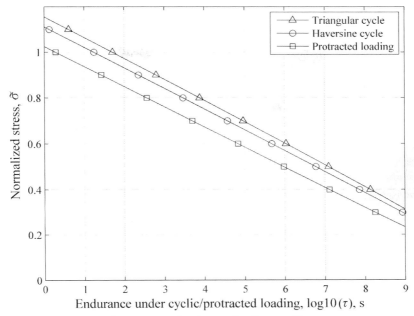

Figure 5.3 Fatigue performance under distinct loading profiles.

- For the haversine cycle

$$\tau_{Ns} \equiv N_s \vartheta = \frac{\tau_0 \exp\left(\dfrac{2\varpi_0 - \alpha_0 \widetilde{\sigma}_{max}(1+r)}{2\widetilde{T}}\right)}{I_0\left(\dfrac{\alpha_0 \widetilde{\sigma}_{max}(1-r)}{2\widetilde{T}}\right)} \tag{5.52}$$

Overall, a cyclic loading presentation via integral (5.42) is sufficient to accurately characterize serviceability of a composite structure undergoing cyclic loading, reflecting the influence of the parameters of this loading.

Figure 5.3 illustrates this opportunity for PMC fatigue performance under three conventional long-term test loading profiles, triangular and haversine waveforms along with protracted unaltered stress rupture loading. Ambient temperature for all these load cases is kept normal in the given illustration to focus on the influence of the minor distinction of the given loading profiles.

The acquired result is computed for the stress ratio $r=0$ that produces the maximal possible discrepancy of fatigue performance under cyclic and protracted loadings. Changing it towards the stress ratio $r \rightarrow 1$ will lead to coalescence of the three lines pertinent to the different loading profiles.

As the influence of loading parameters is countable, results of monotonic loading to failure properly adjusted to match the parameters of the cyclic

loading can be incorporated into the acquired fatigue test data. In particular, expression (5.35) is used for this adaptation.

> As the influence of loading parameters is countable, results of monotonic loading to failure, being properly adjusted to match the parameters of cyclic loading, can be incorporated into the acquired fatigue test data.

Along with pulse cyclic loading, reverse cycling represents another conventional fatigue test option pertinent to operation of ship hull structures. As usually observed in test results, the reverse character of cyclic loading insignificantly affects the fatigue life of structural PMCs compared to that for pulse loading. This unobvious occurrence is probably due to the following.

In general, reversed cycling is associated with either tension–compression, flexural sagging-hogging, or altered in-plane or interlaminate shear loading. Different constituents of a PMC laminate (fiber reinforcemen, polymer matrix, or a combination of these two) are responsible for bearing the distinct phases of the reversed loadings. Due to this, excluding a shear load case, there is typically considerable disparity between the absolute values of ultimate strength for the two phases of reverse loading, i.e., $|S_+| \neq |S_-|$.

Hence, the equivalence of peak forces $|F_{max+}| = |F_{max-}|$ usually maintained over reverse cycling will result in disparity of the normalized peak stresses $|\widetilde{\sigma}_{max+}| \neq |\widetilde{\sigma}_{max-}|$. This disparity engenders prevailing of one phase of the reverse cycle over the other in terms of fatigue deterioration of the load-bearing material components subjected to those distinct phases of the loading exposure.

A measure of interaction or interference of the material's constituents responsible for withstanding either phase of the reverse cycling could vary depending on the chemical, physical, and structural traits of the constituents and their interrelation. Meanwhile, there should be two conceivable extremes in the load-sharing, one that is associated with absolute independence of the load-bearing constituents and another with full engagement of the constituents of both groups in the shared load-bearing.

For the first occasion, the material's lifetime under reverse cycling would be roughly doubled, compared to that under pulse cycling with the same frequency and peak stress. This is because each load-bearing constituent is actually carrying its load share for just a half loading cycle, alternating with the rest under other semi-cycles, free of the loading exposure. Using the same logic, the second case should result in a lifetime that is about that of pulse cyclic loading.

Combining these two main possibilities, we may anticipate that actual endurance τ_{Nr} of a PMC undergoing reverse cyclic loading is within a range

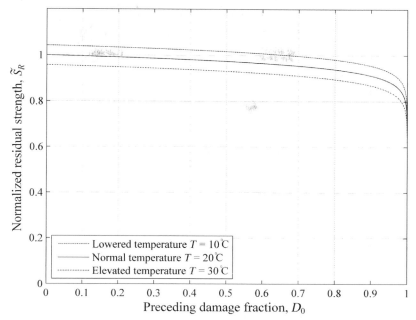

Figure 5.2 Residual strength versus preliminary accumulated damage.

batch articles that do not undergo preliminary loading, due to minor deviation of actual testing conditions from the statutory norm.

Given that the residual strength testing is typically run for large test articles and structure prototypes, it is not always feasible to maintain a temperature-controlled environment during test execution. Because of this, the ambient temperature may readily exceed the range intrinsic to the statutory testing standards, $10 \leq T \leq 30$ °C. The graph in Figure 5.2 illustrates the influence of the expanded temperature variation $\triangle T = \pm 10$ °C on the magnitude of residual strength.

A possible variance of the loading rate, e.g., that due to different rigidity of tested structures, could further extend deviation of the residual strength.

Overall, it should be admitted that experimental determination of residual strength of a composite/hybrid structure may be as informative as that for testing of a metal structure only in the event that all significant testing parameters are properly counted. Otherwise, residual strength testing may produce a quite misleading result.

For this instance, integral (5.28) is presented as

$$D_F = A^{-1} \int_0^{\vartheta_R} \exp\left(\frac{\alpha R}{\widetilde{T}} t\right) dt + D_0 \qquad (5.37)$$

and the solution is sought for $D_F = 1$, as

$$\frac{\widetilde{T}}{A\alpha R}\left(\exp\left(\frac{\alpha S}{\widetilde{T}}\right) - 1\right) = 1 - D_0 \qquad (5.38)$$

Analogously to (5.30)–(5.35), expression (5.38) can be transformed to

$$\widetilde{S}_R = \widetilde{S}_{R1} + \frac{\ln(1 - D_0)}{\alpha_0} \qquad (5.39)$$

where $\widetilde{S}_{R1}$ is a portion of the residual-to-ultimate strength ratio determined similarly to (5.35), as a function of the loading conditions inherent to the monotonic loading of the second test round, as

$$\widetilde{S}_{R1} = \frac{\widetilde{T}_R}{\alpha_0 - \widetilde{T}_R}\left(\alpha_0 + \varpi_0\left(\frac{1}{\widetilde{T}_R} - 1\right) - \ln\left(\widetilde{\vartheta}_R\right) - \ln\left(\widetilde{T}_R\right) - 1\right) \qquad (5.40)$$

For standard loading conditions, as temperature and length of loading are $\widetilde{T}_R = 1; \widetilde{\vartheta}_R = 1$, relation (5.39) is

$$\widetilde{S}_R = 1 + \frac{\ln(1 - D_0)}{\alpha_0} \qquad (5.41)$$

The graph in Figure 5.2 reflects expression (5.39) as a function of the preceding damage fraction $0 < D_0 < 1$, obtained during the first preliminary round of fatigue test loading.

As can be seen from the graph, the preceding damage fraction up to $\sim D_0 \leq 0.8$ only marginally affects residual strength. At a normal temperature, it corresponds to values $S_R \geq 0.94S$ of the anticipated ultimate strength S as determined under standard testing conditions.

In other words, a minor reduction of effective strength, about 6% of the anticipated original ultimate strength S, corresponds to roughly 20% of the material residual load-bearing capability. This analytical result is closely in line with the well-known, though not obvious, phenomenon usually observed during residual strength testing of composite and hybrid structures. In fact, residual strength of PMC test articles may even exceed the experimentally determined ultimate strength of analogues, the same